Current Applications of Engineered Wood

Edited by Jun Zhang

Published in London, United Kingdom

Current Applications of Engineered Wood
http://dx.doi.org/10.5772/intechopen.107739
Edited by Jun Zhang

Contributors
Ahmed Mohamed, Mihaela Tanase-Opedal, Masuod Bayat, Razak Wahab, Mohamad Saiful Sulaiman, Johari Zainuddin, Hashim W. Samsi, Sofiyah Mohd Razali, Ellisha Illing, Abdul Fattah Ab Razak, Yan Xia, Lu Jia, Jun Zhang

First published in London, United Kingdom, 2023 by IntechOpen
IntechOpen is the global imprint of INTECHOPEN LIMITED, registered in England and Wales, registration number: 11086078, 5 Princes Gate Court, London, SW7 2QJ, United Kingdom
Printed in Croatia

British Library Cataloguing-in-Publication Data
A catalogue record for this book is available from the British Library

Additional hard and PDF copies can be obtained from orders@intechopen.com

Current Applications of Engineered Wood
Edited by Jun Zhang
p. cm.

This title is part of the Industrial Engineering and Management Book Series, Volume 1
Topic: Production Engineering
Series Editor: Fausto Pedro Garcia Marquez
Topic Editor: Orhan Korhan

Print ISBN 978-1-83769-162-3
Online ISBN 978-1-83769-163-0
eBook (PDF) ISBN 978-1-83769-164-7
ISSN 3029-0511

For EU product safety concerns:
IN TECH d.o.o., Prolaz Marije Krucifikse Kozulić 3, 51000 Rijeka, Croatia,
info@intechopen.com or visit our website at intechopen.com.

IntechOpen Book Series

Industrial Engineering and Management

Volume 1

Aims and Scope of the Series

Industrial Engineering and Management (IEM) is a discipline that focuses on optimizing complex processes and systems within various industries. It involves the integration of engineering, business, economics, mathematics, and behavioral sciences to improve efficiency, productivity, quality, and overall performance in organizations. Key aspects of Industrial Engineering and Management include: Process Optimization; System Analysis and Design; Quality Control and Management; Supply Chain Management; Operations Management; Human Factors and Ergonomics; Project Management; Cost Analysis and Financial Management; Decision Analysis.

Overall, Industrial Engineering and Management aims to optimize resources, improve processes, enhance productivity, and ensure the effective and efficient utilization of all elements involved in the production or delivery of goods and services. It is crucial in today's competitive business environment for organizations to stay efficient and competitive.

Production Engineering and Operational Excellence are fields of study and practices that focus on optimizing and improving the manufacturing and production processes within an organization. It combines principles from engineering, management, and operational strategies to enhance productivity, efficiency, quality, safety, and sustainability in the production of goods and services.

Here are the key components of Production Engineering and Operational Excellence: Process Optimization; Operational Excellence; Manufacturing Systems Design; Quality Management; Supply Chain Optimization; Production Planning and Scheduling; Automation and Technology Integration; Health, Safety, and Environmental Management; Cost Management; Performance Measurement and Key Performance Indicators (KPIs); Continuous Improvement and Innovation.

Production Engineering and Operational Excellence are crucial for organizations aiming to stay competitive in the global market by achieving high levels of efficiency, quality, and customer satisfaction while optimizing resources and minimizing waste. It is a multidisciplinary approach that encompasses engineering principles, management strategies, and the effective use of technology to drive operational success.

Meet the Series Editor

Fausto Pedro Garcia Marquez is a Full Professor at UCLM, Spain, with accreditation since 2013. He also holds the position of Honorary Senior Research Fellow at Birmingham University, UK, and serves as a Lecturer at the Postgraduate European Institute. In addition to these roles, Fausto has experience as a Senior Manager at Accenture from 2013 to 2014. He earned his European Ph.D. with the highest distinction. Throughout his career, Fausto has received numerous awards and honors. These include the Nominate Prize (2022), Gran Maestre (2022), Grand Prize (2021), Runner Prize (2020), and Advancement Prize (2018), as well as Runner (2015), Advancement (2013), and Silver (2012) by the International Society of Management Science and Engineering Management (ICMSEM). He was also the recipient of the First International Business Ideas Competition 2017 Award. Fausto's contributions extend to academic publishing, with over 242 papers to his name. Notably, his work has been recognized in journals like "Applied Energy" (Q1, IF 9.746, Best Paper 2020) and "Renewable Energy" (Q1, IF 8.001, Best Paper 2014). His affiliations include the editorial and authorship roles in more than 50 books, with publications through respected publishers such as Elsevier, Springer, Pearson, Mc-GrawHill, IntechOpen, IGI, Marcombo, and AlfaOmega. He has authored over 100 international chapters and holds 6 patents. Fausto serves as the Editor of 5 International Journals and is a Committee Member for more than 70 International Conferences. His research portfolio encompasses being the Principal Investigator in 4 European Projects, 8 National Projects, and participating in over 150 projects involving universities and companies. His areas of expertise and research interests span Artificial Intelligence, Maintenance, Management, Renewable Energy, Transport, Advanced Analytics, and Data Science. Fausto is a recognized Expert in the European Union in AI4People (EISMD) and ESF. He also serves as the Director of www.ingeniumgroup.eu, holds the status of Senior Member at IEEE since 2021, and has been honored as an Honorary Member of the Research Council of the Indian Institute of Finance since 2021. Fausto is also the Committee Chair of The International Society for Management Science and Engineering Management (ISMSEM) since 2020.

Meet the Volume Editor

Dr. Jun Zhang is an associate professor at Southwest Forestry University, China, mainly engaged in research on biomass polymer materials and wood composite materials. He is a top young talent in Yunnan Province and a visiting scholar at Kyoto University, Japan; University of Lorraine, France; and University Putra, Malaysia. He has published more than forty papers in international journals such as *Carbohydrate Polymers*, *Industrial Crops and Products*, *International Journal of Adhesion and Adhesives*, and *Progress in Organic Coatings*. He is also a guest editor for *Polymers*.

Contents

Preface

Wood engineering refers to the use of modern engineering technology to process small-sized wood or wood chips into large-sized and high-strength wood component products through bonding and finger bonding. Engineering wood products enables the effective utilization of wood resources. Compared to construction solid wood, the main advantages of engineered wood products lie in their stability and greater structural strength. This means that engineering wood materials can be used in many places where steel is used in construction projects. This book provides a detailed introduction to the types, applications, modifications, mechanical performance testing standards, anti-corrosion modifications, and fungal degradation of engineered wood. It introduces the various types and applications of engineering wood and modified engineering wood. We would like to thank IntechOpen for their support. I am honored to have had the opportunity to edit this book.

Jun Zhang
Yunnan Provincial Key Laboratory
of Wood Adhesives and Glued Products,
Southwest Forestry University,
Kunming, People's Republic of China

Chapter 1

Introductory Chapter: Engineering Wood Review

Jun Zhang

1. Introduction

Forests exist in four major climatic zones (boreal, temperate, subtropical, and tropical) (**Figure 1**). According to the Global Forest Resources Assessment (FRA) report published by the Food and Agriculture Organization of the United Nations (FAO) in 2020, the total area of forests globally amounted to 4.06 billion hectares, representing 31% of the total land area [1]. Meanwhile, the total global area of planted forests is estimated to be 294 million hectares, accounting for 7% of the world's forest area. Asia has the largest area of planted forests with 135.23 million hectares, accounting for 46% of the total global planted forest area, followed by Europe, North and Central America, South America, Africa, and Oceania.

Planted forests are usually defined as forests consisting primarily of planted and/or intentionally seeded trees. Planted forests can provide benefits for traditional timber and fiber production, economic development, and employment in rural areas [2], while they can serve as a key means of combating climate change, restoring degraded land, and maintaining sustainable ecosystems in the short to medium term [3, 4]. In a broader geographic and economic context, well-managed planted forests contribute to sustainable development toward a forest-based circular bioeconomy and healthy ecosystems [5].

Wood is one of the traditional materials used in construction applications, and there is a wide range of engineered wood products (EWPs) available for construction, from sawn lumber to structural lumber. Light-frame systems are the most common type of wood-frame construction, using EWPs such as dimension lumber, placed at regular intervals and fastened together to form floor, wall, stair and roof members. Due to cost advantages, timeliness, and convenience, light-frame wood construction is commonly used in single-family homes, multi-unit dwellings, commercial buildings, and light industrial buildings. Construction costs and lead times are lower than traditional methods because more and more structural components are prefabricated in factories and often shipped to the jobsite along with plumbing fixtures, electrical systems, paints, flooring accessories, and other materials.

The development of wood-frame construction has been uneven globally, with major concentrations in North America, Australia, Japan, and some countries in South-East Asia. In developing countries, although wood is still considered a typical building material, there are a number of constraints that hinder the development of EWPs, such as consumer perception bias, which is usually associated with deforestation; high costs, which are higher when EWPs or more modern building systems are used; a lack of professional builders, who are accustomed to masonry and concrete buildings; and a lack of special regulations and standards. Nevertheless, some changes

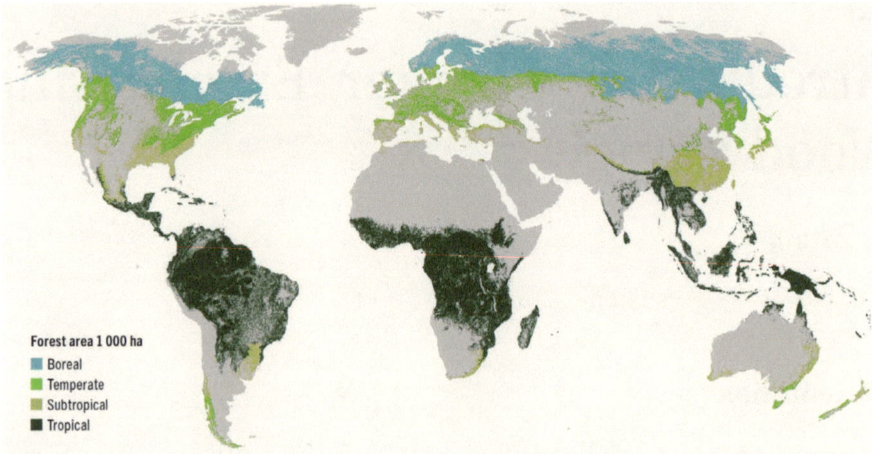

Figure 1.
Global forest distribution by climatic domain [1].

can be observed, especially in some developing countries with high forest cover. In Brazil, for example, the number of companies producing EWPs continues to grow, and the strong links between civil engineering and forestry have resulted in more wood-frame construction.

2. Types of EWPs

EWPs are a man-made composite material made from hardwoods and softwoods. There is a wide variety of EWPs with different manufacturing processes and applications. Examples of EWPs include particleboard, plywood, fiberboard, oriented strand board (OSB), laminated veneer lumber (LVL), glued laminated timber (GLT), and cross-laminated timber (CLT) (**Figure 2**).

2.1 Particleboard

Particleboard is a man-made board made of wood or other lignocellulosic materials made of scraps, applied adhesive, and then glued under the action of heat and pressure; the physical photograph of particleboard is shown as **Figure 3**. As the population grows, so does the market demand for particleboard, and over-exploitation of forest resources has led to a shortage of wood supply in most developing countries. The scarcity of timber resources has limited the development of the particleboard industry; therefore, in addition to timber by-products, some plant raw materials (e.g., bagasse, bamboo, bark, rice husk, etc.) are gradually being used. Bekalo and Reinhardt [6] investigated the process and properties of particleboard prepared from coffee husks. Coffee husk is a coffee processing residue and currently, coffee husk is rarely utilized in Germany and is usually incinerated or landfilled or poured into river water. Akinyemi et al. [7] prepared composite particle boards from waste materials such as corn cobs and wood chips. The effect of waste dosage on the physical and mechanical properties of particleboards was investigated by fixing the volume of adhesive. The results showed that the higher the composition of corn cobs, the faster the boards were saturated by water. The best physical properties were obtained at 50% corn cob

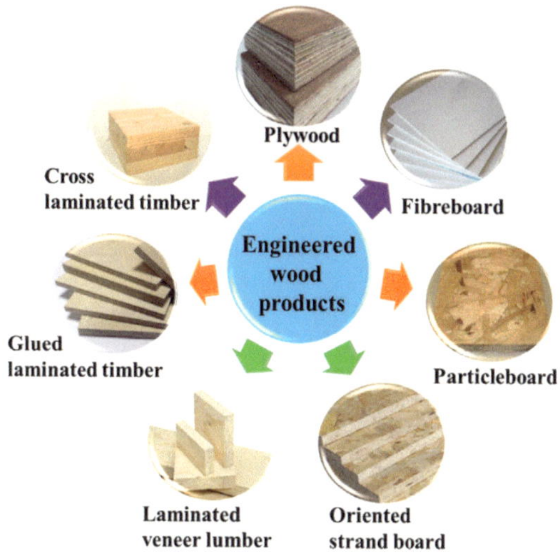

Figure 2.
Types of EWPs.

Figure 3.
Physical photograph of particleboard.

composition and the worst at 100%. Nikvash et al. [8] investigated the properties of three crop processing residues, bagasse, rape straw, and industrial hemp straw, for the preparation of 3-layer structured particleboard. Bagasse was purchased from Iran and rape straw and industrial hemp straw were purchased from Germany. Bagasse, rape straw, and industrial hemp straw were processed into 6 mm long shavings and dried to a moisture content of 3–4%. The surface layer shavings were wood shavings made in Germany, and the core layer shavings consisted of bagasse, rape straw, and industrial hemp straw mixed with wood shavings according to a certain proportion, respectively.

2.2 Plywood

Plywood is generally made of rotary cut veneer or planed thin wood to adjacent layers of veneer fiber direction perpendicular to the group of blanks by the adhesive gluing into a multi-layer wood-based composite material, with a small

Figure 4.
Physical photograph of plywood.

coefficient of deformation, excellent mechanical properties, etc. are widely used in construction, packaging, furniture, flooring, car, and shipbuilding industries, the physical photograph of plywood is shown as **Figure 4**. Over the years, plywood has been one of the leading products in China's wood-based panel industry. With the technological progress and industrial restructuring, the development of plywood industry has entered the key stage of transformation and upgrading, and value-added function has become one of the important ways to increase the added value of plywood, expand its application areas, and enhance the competitiveness of plywood products.

The flammability of ordinary plywood has limited its application in many fields. At present, flame retardant plywood production methods are mainly immersion method, veneer lamination composite method, and surface coating method. Among them, flame retardant plywood is most commonly prepared by veneer impregnation process (**Figure 5**), which is mainly to impregnate veneer or plywood with flame retardant components by pressurized (or atmospheric pressure) method. The current research mainly focuses on the development of new environmentally friendly flame retardant with high impregnation efficiency, good flame retardant effect, small impact on mechanical properties and not easy to precipitate. In addition, in order to meet the more demanding practical application environment, the development of flame retardant multifunctional (aldehyde reduction, low smoke, mold, moisture,

Figure 5.
Preparation of flame retardant plywood by impregnation method.

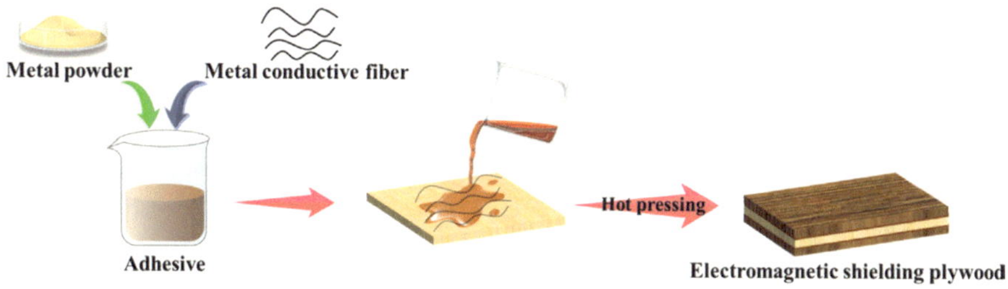

Figure 6.
Preparation of electromagnetic shielding plywood.

antibacterial, etc.) plywood is of great significance [9]. Using ammonium dicyanide, phosphoric acid, magnesium sulfate, boric acid (BX), and other compound treatment of plywood, heat release is significantly reduced by 91.9%, and smoke release is reduced by 76.8%. It has a certain anticorrosive and anti-mold function, while the formaldehyde release of the plywood was reduced by 82.3%.

In addition, electromagnetic shielding plywood is a veneer and electromagnetic shielding materials with electromagnetic shielding material using stacking, mixing, and flexible pressurization and other methods of preparation with the electromagnetic shielding effect of wood-based composite materials. Copper powder, nickel powder, graphite powder, and other conductive powder and iron fiber, copper fiber and other metal conductive fibers added to the adhesive can be used to prepare electromagnetic shielding plywood (**Figure 6**). The metallic copper fibers within a certain size range can effectively improve the electromagnetic shielding effect by increasing the amount of fiber coating and glue coating, which has some practical value. Increasing the amount of conductive material coating can improve the electromagnetic shielding efficiency but is not conducive to the strength of the glue. Metal conductive fiber is more conducive to improving electromagnetic shielding performance than conductive powder under the same amount of conductive material.

Ordinary plywood is susceptible to insect and fungal attack and decay, and its service life and scene are limited. After anticorrosive and anti-insect treatment, plywood has certain anticorrosive, anti-insect, and anti-mold effects, which in turn extends the service life of plywood. The use of impregnation method of horsetail pine and poplar veneer preservative treatment with ammolysis alkylamine copper, borate and different additives compound as preservative, after phenolic or urea-formaldehyde glue gluing can be obtained after a good anticorrosive effect of plywood. It was found that the average drug loading capacity of sound brewing ammonia-soluble alkyl turned horsetail pine and poplar wood preservation plywood was the highest, reaching 7.80 and 9.10 kg/m^{3}, respectively, the average drug loading capacity of adhesive ammonia-soluble copper vanillylamine horsetail pine and poplar wood preservation plywood was 4.21 and 4.53 kg/m^{3}, respectively, and that of UF adhesive BX poplar wood preservation plywood was 4.96 kg/m^{3}, respectively. The average boron retention rate of the phenolic adhesive glyoxal/propanetriol and BX compounded horsetail pine plywood and the phenolic adhesive glyoxal/propanetriol and borax (BA) compounded poplar preservative plywood were 45.52% and 49.38%, respectively, and the preservative plywood produced under the most favorable conditions could reach the strong corrosion-resistant grade.

2.3 Fiberboard

Fiberboard is an artificial panel made from wood fibers or other vegetal fibers, cured by hot pressing under the bonding action of adhesives; the physical photograph of the fiberboard is shown in **Figure 7**. Fiberboard has excellent comprehensive performance and is an important part of packaging, indoor furniture, and decorative materials. However, it is easy to burn and produces smoke and toxic gases when burning, which may lead to fire, and the toxic smoke will cause secondary injury to the human body in the fire, which will bring serious harm to people's living environment as well as personal health and property safety. Therefore, it is necessary to choose safe and harmless green flame retardant to modify fiberboard.

Inorganic flame retardants have the advantages of wide source, low price, environmental protection and safety, good flame retardant performance, and small toxic side effects when they play a flame retardant role, etc. They are the most widely used flame retardants at present and gradually become a hot spot of flame retardant research. Inorganic flame retardants mainly include boron flame retardants, phosphorus and nitrogen flame retardants, metal hydroxide flame retardants, and metal oxide flame retardants. Two borates such as BX and BA are the most commonly used flame retardants in the boron family of flame retardants, which have the advantages of low toxicity to humans and environmental friendliness [10]. Borates are widely used in fire protection because they reduce flame propagation [11]. In addition, the combined use of BX and BA has a synergistic flame-retardant effect [12]. However, inorganic flame retardants use the process of moisture absorption and loss and other shortcomings, and a single inorganic flame retardant is difficult to meet the application requirements, the use of the process is often used in a variety of composite, to obtain excellent performance of the flame retardant smoke suppressant.

In order to overcome the shortcomings of inorganic flame retardants, such as moisture absorption and loss, researchers have developed organic flame retardants on the basis of inorganic flame retardants. Organic flame retardants have good compatibility with the base material fiber and excellent anti-loss performance. Organic flame retardants mainly include organophosphorus and nitrogen, organophosphorus and boron, organophosphorus and nitrogen and boron. Due to the high production cost and unstable performance of organic flame retardants and other shortcomings affecting its application, the application of organic flame retardants in fiberboard research

Figure 7.
Physical photograph of fiberboard.

has rarely been reported, and more organic flame retardants and inorganic flame retardants composite, the preparation of better performance of the flame retardant.

2.4 Oriented strand board

OSB is a kind of wood structural board made from small diameter timber, mesquite timber, wood core, and other raw materials, after slicing, drying, gluing, oriented paving, hot press molding, and other processes, with high strength, high bending strength, good nail grip, less glue, less formaldehyde emission, anticorrosive, anti-moth-eaten, anti-deformation, heat insulation, sound insulation properties, etc. The physical photograph of OSB is shown in **Figure 8**. OSB can be used to replace structural plywood in applications. OSB can replace structural plywood in applications, but compared with structural plywood, the process of obtaining structural units through the planing process makes it less demanding on raw materials, so the raw material sources are more extensive. At the same time, in the OSB manufacturing process, the wood utilization rate is high, up to more than 80%, which can efficiently utilize the wood and achieve the purpose of "inferior wood, better use". With its excellent overall performance, OSB is used as a building panel in Europe and the United States for flooring, wall panels and structural support materials, and is now also used in the wood packaging sector, mainly for packaging pallets and container floors. In China, OSB is mainly used for furniture and interior decoration, commonly used as door frames, shelves, and interior wall panels.

The main research focuses on the theoretical and experimental studies of various processes in the OSB preparation process, including the influence of particle lay-up on the mechanical properties of the boards, the study of OSB lay-up structure and sectional density, the theory of bonding interface between particles and adhesive in OSB, the modeling study of OSB processing and expansion, and the influence of adhesive and its dosage on the overall performance of OSB [13, 14]. In addition, it also includes the modification of OSB by adding borate, the improvement and optimization of hot pressing process parameters, the optimization of directional paving process, the research and development of OSB adhesive for broadleaf timber, and the improvement of particle production process, etc.

Comprehensive analysis of the current research situation can be found; the current OSB research has achieved great results. Researchers on OSB manufacturing

Figure 8.
Physical photograph of OSB.

process of raw materials and basic process parameters for a more in-depth study; but at the same time should also be noted that, compared with other wood-based panel products, OSB research is still somewhat insufficient but also need to continue to strengthen the research on the theoretical and practical aspects of the two.

2.5 Cross-laminated timber

In recent years, CLT has gained popularity in Europe and is gradually gaining interest in the rest of the world due to its strength, appearance, versatility and sustainability. The material consists of sawn, glued, and layered wood panels where each layer is perpendicular to the previous one. The layers of wood are joined at a perpendicular angle, allowing the structural stiffness of the panel to be obtained in both directions, similar to plywood, but with thicker components. This gives the panel great tensile and compressive strength. A physical drawing of CLT is shown in **Figure 9**, and the dimensions of CLT for different applications are shown in **Figure 10**.

CLT is a sustainable material because it is composed of wood, a renewable resource (often from reforestation), and does not require the burning of fossil fuels during its production. It has been used for infrastructure and support on large construction sites, as a form of concrete bridge, and even as a foundation for tractors in unstable terrain during dam construction. Due to its interesting appearance and structural strength, its potential in smaller structures has been noted. Currently, there are even skyscrapers built using CLT parts.

In fact, CLT is not in competition with the existing timber building sector, with its focus on linear timber elements, but a direct competitor of mineral-based solid building materials. This position is expected to be further strengthened. This is due to the fact that local timber species can be sustainably utilized to the benefit of all regions of the world.

When designing CLT structures, it is necessary to consider not only specific knowledge about CLT and joint design but also the whole structure, utilizing integrated knowledge and interdisciplinary thinking.

2.6 Classification of adhesives for EWPs

Wood adhesives can be broadly categorized as petroleum-based or natural adhesives. Petroleum-based adhesives can be further categorized as thermosets,

Figure 9.
Physical diagram of CLT.

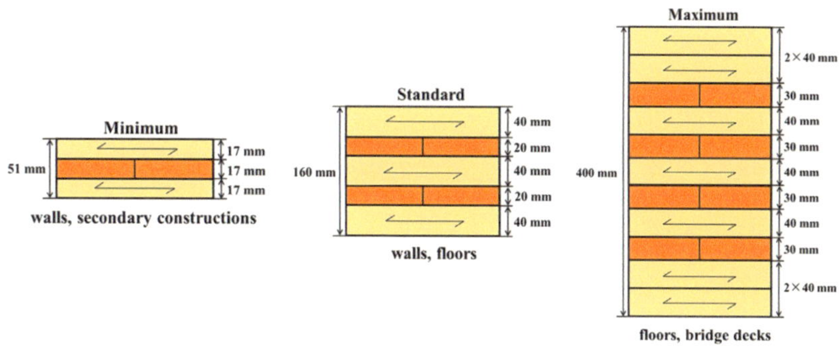

Figure 10.
Dimension diagram of CLT for different purposes.

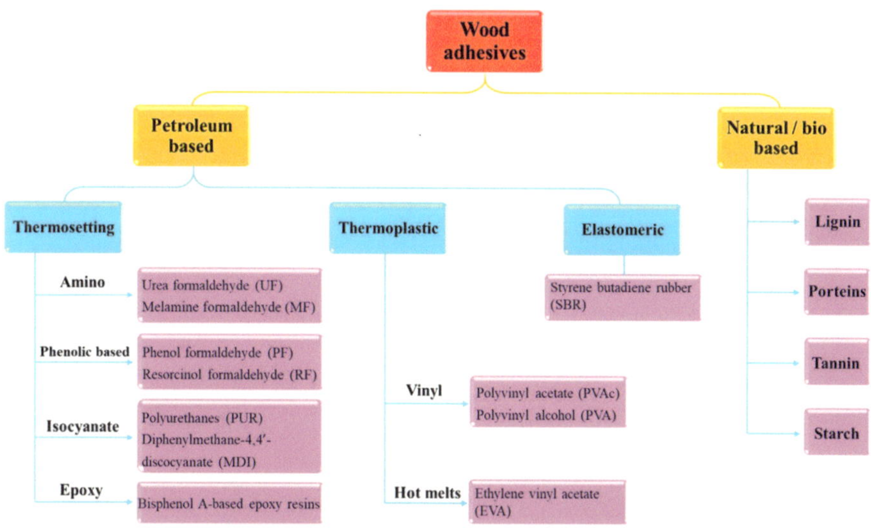

Figure 11.
The classification of wood adhesives.

thermoplastics, and elastomers. Natural or bio-based adhesives can be derived from four main sources, namely lignin, proteins, starch, and tannin. **Figure 11** shows the classification of wood adhesives and their subclassifications.

Commonly used petroleum-based adhesives include UF, phenol-formaldehyde (PF), melamine formaldehyde (MF), resorcinol formaldehyde (RF), and isocyanate-based adhesives [15]. However, growing environmental concerns and the increasing depletion of petroleum-based resources have put pressure on the wood composites industry to develop environmentally friendly adhesives using renewable resources [16].

Author details

Jun Zhang
Yunnan Provincial Key Laboratory of Wood Adhesives and Glued Products,
Southwest Forestry University, Kunming, People's Republics of China

*Address all correspondence to: zj8101274@163.com

References

[1] Agriculture Organization of the United Nations (FAO), Global Forest Resources Assessment 2022: Main Report, FAO, Italy, 2022. Available from: https://www.fao.org/3/cb9360en/online/src/html/deforestation-land-degradation.html

[2] Malkamäki A, D'Amato D, Hogarth NJ, et al. A systematic review of the socio-economic impacts of large-scale tree plantations, worldwide. Global Environmental Change. 2018;**53**:90-103

[3] Korhonen J, Nepal P, Prestemon JP, et al. Projecting global and regional outlooks for planted forests under the shared socio-economic pathways. New Forests. 2021;**52**(2):197-216

[4] IPCC (Intergovernmental Panel of Climate Change). Climate change and land summary report. 2019. Available from: https://ipcc.ch/report/srccl

[5] Bauhus J, Meer PJ, Kanninen M. Ecosystem Goods and Services from Plantation Forests. United Kingdom: Earthscan; 2010

[6] Bekalo SA, Reinhardt HW. Fibers of coffee husk and hulls for the production of particleboard. Materials and Structures. 2010;**43**:1049-1060

[7] Akinyemi AB, Afolayan JO, Oluwatobi EO. Some properties of composite corn cob and sawdust particle boards. Construction Materials and Technology. 2016;**127**:436-441

[8] Nikvash N, Kraft R, Kharazipour A, et al. Comparative properties of bagasse, canola and hemp particle boards. European Journal of Wood and Wood Products. 2010;**68**:323-327

[9] Xu F, Zhang H, Wu J. Synergistic catalytic flame retardant effect of zirconium phosphate on the poplar plywood. Construction Materials and Technology. 2021;**290**:123208

[10] Nagieb ZA, Nassar MA, El-Meligy MG. Effect of addition of boric acid and borax on fire-retardant and mechanical properties of urea formaldehyde saw dust composites. International Journal of Carbohydrate Chemistry. 2011

[11] Rejeesh CR, Saju KK. Effect of chemical treatment on fire-retardant properties of medium density coir fiber boards. Wood and Fiber Science. 2017;**49**(3):332-337

[12] Özdemir F, Tutus A. Effects of fire retardants on the combustion behavior of high-density fiberboard. Bioresources. 2013;**8**(2):1665-1674

[13] Lee SH, Ashaari Z, Jamaludin FR, et al. Physico-mechanical properties of particleboard made from heat-treated rubberwood particles. European Journal of Wood and Wood Products. 2017;**75**:655-658

[14] Klímek P, Wimmer R, Meinlschmidt P, et al. Utilizing Miscanthus stalks as raw material for particleboards. Industrial Crops and Products. 2018;**111**:270-276

[15] Nuryawan A, Alamsyah EM. A review of isocyanate wood adhesive: a case study in Indonesia. Adhesive Bonding: Science, Technology and Applications. 2017:73-90

[16] Zhang X, Zhu Y, Yu Y, et al. Improve performance of soy flour-based adhesive with a lignin-based resin. Polymers. 2017;**9**(7):261

Chapter 2

Types of Engineered Wood and Their Uses

Masuod Bayat

Abstract

Engineering wood, also known as composite wood or manufactured wood, is a versatile and sustainable material that has gained significant popularity in various engineering applications. Engineering wood is created by combining natural wood fibers or particles with adhesives and other additives to enhance its strength, durability, and dimensional stability. The resulting material exhibits improved properties compared to traditional solid wood, such as a higher strength-to-weight ratio, resistance to moisture and pests, and reduced warping or shrinking. One of the primary applications of engineering wood is in the construction industry. Another significant application of engineering wood is in the furniture industry. It can be molded into various shapes and sizes, allowing for intricate designs while maintaining structural integrity. Engineered wood products like plywood and medium-density fiberboard (MDF) are commonly used for manufacturing cabinets, tables, chairs, and other furniture pieces. Engineering wood also finds application in the automotive sector. The versatility of engineering wood extends beyond the construction, furniture, and automotive industries. In conclusion, engineering wood offers a wide range of applications across various industries due to its enhanced properties compared to solid wood. Its use in construction provides durable structural elements while reducing environmental impact.

Keywords: engineering wood, environment, engineering wood product, engineering wood types, engineering wood applications

1. Introduction

Wood has been used by humans in various fields since ancient times, but in recent years, environmental problems have made governments and global health officials think of a solution to minimize ecological pollutants Above all, he thought about preserving the forests. Wood has living and dead cells, which has caused its properties to be strongly affected by factors such as time, humidity, heat, etc. Wood has relatively good properties against pressure, bending, and stretching. Also, wood has relatively good heat and sound insulation properties. Unlike steel and many other metals, wood is reversible. In addition, wood has a natural and impressive beauty that has caused it to be used in various industries, including furniture production, interior and exterior decoration of buildings, etc. [1].

Wood is one of the most important construction and industrial materials obtained from trees. Wood has unique physical, chemical, and mechanical properties that make it suitable for use in various industries, including construction, furniture production, paper production, etc. Wood is one of the most important natural resources used in various industries such as construction, furniture, and vehicles.

The decrease in the level of forests and natural resources is reaching more acute and worrying stages day by day; So that nowadays the efforts of all countries are toward the optimal use of wood raw material. Wood is one of the oldest building materials and has many advantages as a building material. Therefore, by using natural and recyclable wood products, environmental degradation can be controlled to some extent. Also, reducing energy consumption and costs is very important [2].

However, improper use of forests and neglecting issues related to environmental protection can lead to air pollution, biodiversity loss, environmental degradation, and climate change. To preserve the environment and sustainable use of wood, new approaches have been developed, such as planting sustainable forests, recycling wood, using engineered wood, and using alternative materials such as polymers and metals. Also, new technologies such as wooden buildings with minimal energy consumption and the production of recycled paper help to preserve the environment. The decrease in the level of forests and natural resources is reaching more acute and worrying stages day by day; So today the efforts of all countries have led to the optimal use of wood raw material [3]. However, due to improper harvesting of forests, environmental degradation, biodiversity loss, and climate change, wood use may harm the environment [4]. To preserve the environment and sustainable use of wood resources, new approaches such as planting sustainable forests, wood recycling, using engineered wood, and using substitute materials for natural wood such as polymers and metals have been developed [5].

The use of wood waste as raw materials in the production of products is very important, due to the reduction of harvesting from forests and the preservation of the environment. Also, by recycling wood waste, you can save natural resources and reduce production costs [6].

Engineered wood is wood that has been improved using chemical, physical, and mechanical processes. This type of wood can be used as a substitute for natural wood in many applications such as the construction industry, furniture industry, art industry, etc., by increasing its strength, hardness, dimensional stability, and other mechanical properties [7]. **Figure 1** shows the use of wood from ancient times to today.

Engineering wood, also known as composite wood or manufactured wood, has gained significant recognition and usage in various industries due to its versatility and

Figure 1.
The use of wood from the past to the present [8].

enhanced properties compared to traditional solid wood. This engineered material is created by bonding together multiple layers of wood veneers or fibers with adhesives, resulting in a strong and stable composite product [9]. The development of engineering wood has revolutionized the construction, furniture, and interior design industries, offering an array of benefits and applications.

This introduction aims to provide an overview of engineering wood and explore its wide-ranging applications in different sectors. One of the primary advantages of engineering wood is its improved strength and stability compared to solid wood [10]. By combining different layers of wood, this composite material achieves enhanced structural integrity, making it suitable for demanding applications in construction. Additionally, engineering wood exhibits superior dimensional stability, reducing the risk of warping, twisting, or splitting that can occur with solid wood. Another significant advantage of engineering wood is its environmental sustainability. Many engineered wood products are manufactured using wood fibers sourced from sustainably managed forests, ensuring responsible and eco-friendly production [11].

Additionally, the utilization of engineered wood reduces the demand for solid wood, contributing to the conservation of natural resources. The versatility of engineering wood is evident in its wide range of applications. In the construction industry, it is commonly used for structural components such as beams, columns, and floor systems, where its strength, stability, and durability are key factors. Engineering wood can also be found in furniture manufacturing, enabling the creation of esthetically pleasing and long-lasting pieces. Its consistent quality and resistance to warping make it an ideal choice for cabinetry, tables, chairs, and other furniture items. Furthermore, engineering wood finds its place in interior design, where it is utilized for wall paneling, decorative elements, and various architectural features [12].

By understanding the characteristics and applications of engineering wood, professionals and enthusiasts can make informed decisions when selecting materials for their projects. This knowledge allows for the utilization of engineering wood's unique properties to achieve desired outcomes in terms of strength, durability, sustainability, and esthetics. In the following sections of this article, we will delve deeper into the different types of engineering wood, their specific applications, and the advantages they offer. By exploring these aspects, readers will gain a comprehensive understanding of engineering wood and its potential to transform various industries.

2. What is the application of wood?

2.1 Woods

The main uses of wood are fuel consumption, however, we must remember that wood was the main source of energy and fuel before the year 1850. Currently, wood is seen as a fuel mostly in developing countries, which can be mentioned in African and South American countries.

People who live in industrialized countries and their main source of energy supply is oil should know that their energy supply sources (coal, gas, oil) were created from decomposed forests thousands of years ago. Once consumed, it takes millions of years to be replenished. Some scientists believe that the current forests, with careful management, can be a source of inexhaustible production of firewood and charcoal, which are very cheap [13].

Today, wood is used a lot. Wood is an excellent material for use as fuel. On the other hand, in the construction industry, wood is used to make materials or the body of the building. The use of wood for building construction dates back more than 6000 years ago, and this material is still widely used in the construction industry. On the other hand, some people use wood as a material to decorate the building. Wood is used to make all kinds of tables and furniture, decorative and consumables, and it is considered one of the most widely used and common materials for making partitions, flooring, doors, and windows [14].

In fact, more than 5000 different wooden products are continuously produced. Products of this category have been known for a long time, some of which are: cellulose, wood varnish, and artificial silk [15].

Another increasing use of wood is in the production of compact boards such as plywood, chipboard, and fiberboard. Even the complex systems that made space travel possible require the major use of wood and its products, which can be referred to as a type of cork from oak bark as thermal insulation.

The following are the most common uses of wood:

- Construction of wooden doors and windows
- Making wooden decorative accessories
- Use as fuel
- Making furniture
- Use in building construction
- Create a partition
- Construction of wooden flooring and wall covering
- Use in shipping and boat-building industries
- Making paper and packaging industries

2.2 Wood properties

Wood is a unique and attractive living texture that cannot be compared with any other material. Wood has great resistance and strength and is very flexible. The weight of wood is lighter than other materials and it can be made into different shapes. Another feature of wood is that it has different properties in different ways. You should know that wood is extremely durable. Many wooden items from hundreds or even thousands of years ago are still intact. Although one of the most important negative characteristics of wood is its inability to tolerate moisture. So that the exposure of wood to moisture can cause rotting of the wood texture as well as the growth of insects and fungi in it.

The most attractive feature of wood that makes it very distinctive is its amazing design and color. Woods have different colors and designs, and therefore they can be used to make different materials or different designs [16].

Converting wood into fibers and smaller components (chips) and then recombining these components into a variety of composite products makes the manufactured products have a more uniform resistance in different directions.

In addition to the mentioned cases, the environmental effects brought to the region due to mining often require more time than a human's lifetime to restore. While many trees can be produced during this time [17].

3. Types of engineering wood and their use

Engineered wood, which is also known as industrial wood and composite wood. It is a wood product derived from wood waste and by-products such as sawdust. In fact, these waste woods undergo changes and transformations using heat, glue, and pressure to obtain a suitable and usable replacement for natural wood.

Engineered wood is considered a very prominent development in the production of building materials. Which has many advantages compared to natural wood. Since wood waste is used in the construction of engineered wood. (As a result, there is no need to cut more trees). From the point of view of the environment, they are not only harmful, but they are also friendly and preserve the environment.

The thickness and density of this type of wood can be ordered based on the needs and goals of the project. In addition, engineered wood has high strength. This type of wood can be easily manipulated and cut with simple tools and skills [18].

Wood is one of the most efficient and useful materials that has been available to man since ancient times. Since the beginning of creation, man has used stone and wood as the first building materials. In total, three types of wood are used to make wooden accessories. These three types include soft wood-hard wood-engineering wood. If you take a quick look around you, you will see that most of your things are made of wood, and most of this type of wood is made up of engineered wood. If you do not know what type of engineered wood your bed, desk, chair, and any of these home appliances are, then stay with us. What kind of wood is called engineering wood? The reason for the engineering name of this type of wood is that the production process of such wood is designed precisely. This type of wood is the result of using waste and can be said to be a by-product of wood. The collected wastes are connected utilizing heat and glue. Of course, engineering wood is also known by names such as composite and industrial wood. Due to its very interesting and appropriate structure, engineered wood created a huge change in the construction of wooden materials and decoration [19].

3.1 Types of engineered wood

3.1.1 Laminated board

Several thin layers of wood are placed on top of each other and these thin layers are connected at different angles using glue. As mentioned before, this wood board is formed by stacking thin sheets on top of each other. It is recommended to use an angle of 45 degrees to overlap the thin layers in the construction of these boards; Because the angle of 90 degrees cannot provide enough resistance for multi-layer boards. Therefore, we connect the thin layers at an angle of 45 degrees. This prevents easy bending and twisting of the wood [20]. This type of wood is shown in **Figure 2**.

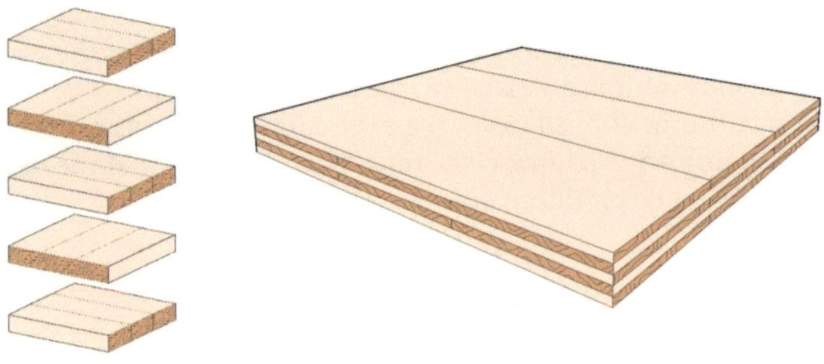

(**a**) single layered structure (**b**) laminated structure.

Figure 2.
Laminated wood [21].

3.1.1.1 Different types of laminated board

- Multi-layer softwood board (made from wood such as beech, mahogany, and birch)
- Multi-layer hardwood board (made from pine, fir, and cedar wood)
- Tropical multilayer board
- Decorative multi-layer board (suitable for construction and decoration design)

3.1.1.2 Applications of laminated board

Laminated boards have a wide range of applications in various industries and sectors. Some common applications include [22, 23]:

1. Furniture: laminated boards are extensively used in the furniture industry for making tables, cabinets, shelves, and other types of furniture. The laminated surface provides durability, resistance to scratches and stains, and an attractive finish.

2. Interior design: laminated boards are popularly used for wall paneling, flooring, and decorative purposes in interior design projects. They offer a wide range of colors, patterns, and textures to enhance the esthetics of residential and commercial spaces.

3. Construction: laminated boards find applications in construction projects for making doors, windows, partitions, and false ceilings. They provide structural strength while offering an appealing appearance.

4. Packaging: laminated boards are used in packaging industries for making boxes, cartons, displays, and other packaging materials. The lamination protects against moisture, dust, and damage during transportation.

5. Signage and advertising: laminated boards are commonly used for creating signage boards, billboards, display stands, and advertising materials due to their durability and ability to withstand outdoor conditions.

6. Automotive industry: laminated boards are utilized in the automotive industry for making interior components such as dashboards, door panels, trims, and headliners. They offer resistance to heat, moisture, and wear while providing a visually appealing finish.

7. Educational materials: laminated boards are often used in educational settings for making whiteboards or blackboards that can be written on with markers or chalk. They provide a smooth writing surface that can be easily cleaned.

8. Exhibition booths: laminated boards are commonly used in exhibition booths or trade show displays as they can be easily customized with graphics or branding elements while being lightweight for easy transportation.

9. Retail displays: laminated boards find applications in retail environments for creating displays, shelves, and fixtures. They offer a clean and professional appearance to showcase products effectively.

3.1.1.3 What are the most important advantages of laminated wood boards?

The main and very important benefits,

1. Strength and endurance: one of the main and very important advantages, the multilayer board has a lot of resistance due to its composition of several layers. It makes optimal use of the large and larger surface of wood by consumers who can easily do many works on the surface of the laminated. Of course, this resistance is distributed throughout its surface.

2. Size and criteria: the boards are very suitable for making wooden items due to the texture of the wood and the way the layers are placed on top of each other in addition to their high strength. Usually, the size of these wooden boards is 180 cm long and 120 cm wide.

3.1.1.4 What are the disadvantages of using multi-layer wood boards?

1. Considering their strength, their price is not very suitable for the consumer.

2. Due to the multi-layered nature of this wood board, it is more difficult to cut. This makes it difficult to cut these wood boards.

3. Dangerous gasses that may be released when cutting this wood board can be harmful to people.

3.1.1.5 What are the most important applications of multi-layer wood boards?

1. Furniture industry

2. Seats and table types

3. Types of wooden shelving

4. Stalling and display of exhibitions

5. Making all kinds of wooden doors

3.1.2 Chipboard (particle board)

Like other types of engineered wood, chipboard is obtained from waste wood such as sawdust and wood chips. These wood wastes are converted into wood chips during a special process, and these wood chips are well mixed with resin or glue during the thermal compression process. The final product is called a chipboard.

Chipboard is a type of engineered wood product that is made from compressed wood chips. It is one of the most widely used types of engineering wood, which is also known as chipboard. If you pay attention to the structure of this wood, you will see wood waste, such as wood chips, in its structure. These wood wastes, which are mostly wood chips, are compressed by heat and pressure and connected by glue or resin. In addition, note that to improve some characteristics of this wood, such as resistance to moisture, fire, or the production of soundproof panels, he made changes and corrections in the boards [24]. An example of chipboard wood is shown in **Figure 3**.

3.1.2.1 Applications of chipboard (chipboard)

Chipboard, also known as particleboard, is a versatile and commonly used engineered wood product. It consists of wood particles or chips bonded together with an adhesive under heat and pressure. Chipboard has a range of applications across various industries, including construction, furniture manufacturing, packaging, and interior design. Here are some common applications of chipboard [26, 27]:

Figure 3.
Chipboard wood [25].

1. Furniture manufacturing: chipboard is extensively used in the production of furniture, including cabinets, shelves, desks, tables, and wardrobes. It serves as a cost-effective alternative to solid wood and can be laminated or veneered to enhance its esthetic appearance.

2. Interior design: chipboard panels are widely employed in interior design applications. They are used as wall paneling, decorative panels, room dividers, and ceiling tiles. Chipboards can be painted, laminated, or covered with decorative veneers or laminates to achieve the desired design esthetic.

3. Flooring: chipboard is used as a substrate for laminate flooring. It provides a stable and durable base for the laminate layer and helps to create a budget-friendly flooring option.

4. Packaging: chipboard is commonly used across various packaging applications. It is used in the production of cardboard boxes, carton packaging, and other types of packaging materials. Its strength and durability make it suitable for protecting and shipping various products.

5. Construction: chipboard finds application in the construction industry as well. It is used for sheathing, subflooring, roof decking, and wall partitions. Its structural strength and affordability make it a preferred choice for these applications.

6. Soundproofing and insulation: the structural properties of chipboard make it useful for soundproofing and insulation purposes. It can be used as an underlayment material to reduce sound transmission or as an insulation material in walls, floors, or roofs.

These are just a few examples of the many applications of chipboard. Its affordability, versatility, and ease of use make it a popular choice across multiple industries where cost-effective and robust wood-based materials are required.

3.1.2.2 Advantages and disadvantages of chipboard

1. Weight: due to the conditions of preparation and compression, this wood has lightweight. If we want to compare this wood with the multi-layer type of engineering wood these wood have a much lower weight. This issue has made it more comfortable in terms of transportation and handling compared to other types of this category of wood.

2. Price: if we want to mention the most important advantage of using this wood, it is undoubtedly its affordability. Big companies like IKEA (is the name of a popular Scandinavian-founded, worldwide furniture store) use chipboard and MDF to make high-quality and low-cost furniture. These woods are mostly used in large pieces such as dining tables, desks, TV tables, and bookshelves.

3. Insulating: it is used in making furniture and internal uses; Because these standard woods are not suitable for parts that are prone to getting wet or in high humidity. This wood is cheaper than miso wood.

4. Nail maintenance: one of the main needs that wood must have is the maintenance of nails and screws. During the investigation and comparisons that have been made, these woods are more maintainable than MDF. Chipboard has a great ability to accept glue and paint.

5. Malleability: the structure of this wood model is such that you can easily change its size using a cutting machine and a drill. Buy chipboard raw or coated with thin sheets of veneer or plastic.

Some of the most important disadvantages of chipboard.

1. It has low moisture resistance. If it is exposed to moisture, its surface will swell and become scaly and its color will change.

2. In general, it can be said that chipboard does not have a very beautiful appearance and is mostly used in cases where the appearance of the work is not of great importance.

3. It has little strength. In general, it can be said that the strength and resistance of the chipboard are low and it does not have a high tolerance against heavy load and pressure.

4. It is not suitable for heavy loads because it cannot bear them.

5. More destructibility than other types of wood

3.1.3 MDF

Medium-density fiberboard, commonly known as MDF, is made by combining sawdust, wood chips, or even organic fiber and pressing them with high pressure. MDF, as a multi-layer board, is widely used as a building material in residential and commercial projects. Perhaps the only disadvantage of MDF is that it is very dense and therefore weighs significantly compared to plywood and chipboard.

Undoubtedly, MDF is one of the most widely used woods used in making construction materials and wooden items. They use MDF to make cabinets, furniture, wall panels, decorative items, and doors; Because MDF is heat and sound insulation. MDF is also used in acoustic enclosures such as speakers or bass speakers. For people who work with MDF, the only problem they get from it is its high weight. Usually, the use of MDF in floor products, such as laminate and parquet floors, is not practical and its use is completely wrong. In the following, we will discuss some points about the use, advantages, and disadvantages of MDF [28]. An example of MDF wood is shown in **Figure 4**.

3.1.3.1 What are the most important advantages of MDF?

1. Insulation: one of the key points of this type of engineered wood is sound insulation. This point has caused the use of this wood in the music industry as well. to prevent excessive vibration. As well as insulation against moisture and dryness.

2. Price: this type of wood allows you to use high quality at a low price.

Figure 4.
MDF wood [29].

3. Resistance: MDF is resistant to termites and other pests due to the use of special chemical processes in its manufacture.

3.1.3.2 What are the disadvantages of using MDF?

1. MDF wood requires a lot of maintenance. If the MDF fills or cracks, repairing it is not an easy task at all.

2. Sometimes it may absorb moisture earlier than natural wood and swell.

3. The chemicals used in the structure of MDF are not very suitable for children.

4. The type of glue used in MDF makes it a little difficult to use some connections such as screws in this material.

5. Although suitable for nail storage, problems may arise at first and the wood is prone to cracking.

6. For People who work with MDF, the sawdust produced from it can cause respiratory problems in them.

3.1.3.3 Applications of MDF

Medium-density fiberboard (MDF) is a versatile material that finds applications in various industries and sectors. Some common applications of MDF include [30, 31]:

1. Furniture and wooden items. Doors and door components: MDF is utilized indoor construction, both as a core material and as a surfacing material for panel doors. It provides stability, durability, and a consistent appearance, making it suitable for interior doors.

2. Cabinet making and interior decoration

3. Flooring: MDF can be used as an underlayment for laminate or engineered wood flooring. Its high density provides stability and helps to reduce noise transmission.

4. Speaker enclosures: the acoustic properties of MDF make it an ideal material for constructing speaker enclosures. Its density helps reduce unwanted vibrations and resonances that can affect sound quality.

5. Art and crafts: MDF is a popular choice for artists and crafters due to its smooth surface, which allows for easy painting, carving, or engraving. It can be used as a canvas or as a base material for creating sculptures, signs, or decorative items.

6. Display fixtures: MDF is frequently used in retail environments to create display fixtures such as shelves, racks, and stands. It can be easily customized to fit specific product requirements and offers a cost-effective alternative to solid wood.

3.1.4 HDF

If you are looking for an economical and reliable alternative to wood and plywood, boards made of wood fiber such as (high-density fiberboard) HDF and MDF are the best choice. Like MDF, HDF (high-density fiberboard) is a composite sheet made of pressed wood particles. HDF is a very thin sheet whose thickness generally varies between 3 and 8 mm. HDF is a man-made wooden product that is produced from the combination of wood chips impregnated with synthetic resins and adhesives. The mixture of these wood and resin particles is subjected to high temperature and pressure and becomes thin sheets with a thickness of less than 1 cm. The wood chips used in HDF are much more homogeneous than MDF and chipboard, which is why it has a higher density than them, around 900 kg/m^3 [32]. HDF has a higher density compared to other types of fiberboard, such as medium-density fiberboard (MDF). This increased density gives HDF superior strength, durability, and resistance to moisture and impact. Due to its high density and strength properties, HDF provides excellent stability and structural integrity. It also has good screw-holding capacity and can be machined with sprecision. Additionally, HDF has low formaldehyde emissions compared to some other wood products. An example of HDF wood is shown in **Figure 5**.

3.1.4.1 Advantages and disadvantages of using HDF wood

1. HDF is a high-priority substrate for multi-layer wood flooring; Because it is hard and solid.

2. Another advantage of HDF wood is its completely smooth surface, and this smooth surface makes it suitable for the production of flooring.

3. This type of engineering wood has very high color acceptability.

4. HDF is also a good choice for use in frames, boxes, shutters, and internal shelves of fitted wardrobes.

Figure 5.
HDF wood [33].

5. In addition to the price, it looks beautiful and is similar to natural wood, which makes this type of wood more popular than natural wood.

6. One of the disadvantages that distinguishes HDF from other types of engineered wood such as chipboard is its high resistance to pests such as termites. The reason for this is the materials and chemical compounds used in the production process of this type of wood.

3.1.4.2 Disadvantages of HDF

1. Like chipboard, this wood is not highly resistant to moisture and water.

2. This wood cannot be used in open spaces due to the high absorption of moisture.

3.1.4.3 Applications of HDF

High-density fiberboard (HDF) is a durable and versatile engineered wood product with several applications across various industries. Here are some common applications of high-density fiberboard (HDF) [34, 35]:

1. Wood furniture industry and door making: HDF is widely used for making doors and furniture due to its resemblance to natural wood. In addition, with the help of HDF, they make very beautiful wooden items.

2. Flooring and wallpaper: the natural and beautiful appearance of this type of wood has made it a suitable choice for people who want to floor and wallpaper their homes.

3. Wall paneling: HDF panels are employed for decorative wall paneling, wainscoting, or other interior wall applications. The smooth and uniform surface of HDF allows for easy installation and finishing, such as painting or applying surface treatments.

4. Interior design and architectural millwork: HDF is a popular choice for interior design applications, such as wall paneling, wainscoting, moldings, and trim. Its dense composition ensures stability and resistance to warping, making it suitable for installations in various environments.

5. Soundproofing and acoustic panels: the dense structure of HDF makes it an excellent material for soundproofing applications. It is commonly used in the construction of acoustic panels and sound barriers to reduce noise transmission and improve acoustics in commercial spaces, studios, theaters, and home theaters.

6. Crafts and DIY projects: HDF's versatility and ease of working make it a popular choice for crafts and do-it-yourself (DIY) projects. It can be easily cut, routed, and shaped to create custom designs for home décor, signage, model making, and other creative applications.

7. Automotive industry: HDF is utilized in the automotive industry for making interior components such as door panels, dashboards, and trim pieces due to its durability and ability to withstand vibrations.

It's important to note that HDF comes in different grades and thicknesses, allowing for a wide range of applications. The specific characteristics of HDF, such as its density, strength, and moisture resistance, can be tailored to meet the requirements of different applications. Overall, high-density fiberboard offers a cost-effective, versatile, and durable solution for various industries. Its applications span from furniture manufacturing to interior design, flooring, packaging, soundproofing, and beyond, making it a valuable material in numerous sectors.

3.2 What are the disadvantages of engineered wood?

Some of the disadvantages of engineered wood include:

1. Very low reparability: engineered wood has a very thin shell that cracks easily. If this happens, engineered wood will be very difficult to repair.

2. Unsuitable for humid and hot environments: against humidity and heat, engineered wood may shrink or expand and lose its beautiful appearance.

3. Artificial look: the pattern on engineered wood is usually overprinted, which is why it feels artificial.

4. Cost: engineered wood is usually more expensive than traditional wood.

5. Lack of natural attractiveness: engineered wood cannot have the natural attractiveness of traditional wood due to its uniform appearance.

3.3 Which engineered wood is better?

However, on the other hand, engineered wood also has advantages as a cost-effective alternative to traditional wood. For example:

1. The most efficient: engineered wood is made from woods that provide the most efficiency, for example, beech wood and soybean wood.

2. Greater resistance to changes in temperature and humidity: engineered wood can withstand more changes in temperature and humidity.

3. Reducing losses: engineered wood reduces losses due to biological effects, fighting insects and reducing the size and weight of wood.

4. Providing side cover: engineered wood can provide a side cover that protects the wood structure against vandalism factors.

5. Improving base construction: engineered wood can be useful for improving base construction for tall buildings.

In general, engineered wood appears to be preferred in environments seeking to reduce losses.

3.4 Is engineered wood a natural material? Is it environmentally friendly?

Therefore, engineered wood is not a material that is born naturally, but because the base material of the tree is used for its production, it can be considered environmentally friendly. However, it should be said that as a raw material, using traditional wood is better, and using engineered wood should only be done if the project requirements support it. Also, for the long-term sustainability of the environment, it is necessary to ensure the restorative materials used for engineered wood, which usually consist of chemical resins [36].

3.5 Why is engineered wood expensive?

In addition, the production of engineered wood costs more than traditional wood products due to more complex processes and advanced technologies. Also, some types of wood used in the construction of engineered wood are of high quality, which requires additional costs for their extraction. In general, engineered wood can be used due to its advantages such as less wood consumption, lower maintenance and installation costs, greater yield from the tree, and more sustainable applications.

3.6 How long does engineered wood last?

The life of engineered wood generally depends on the quality of construction, the type of trees used, environmental conditions, and how to maintain and use it.

However, for high-quality engineered wood, it has a lifespan of at least 30–50 years. Considering that engineered wood is more resistant than traditional wood, it is usually used in domestic and industrial applications to make walls, doors, windows, parquet, and furniture [10].

3.7 How do you protect engineered wood?

You can use the following methods to preserve engineered wood and increase its life: Regular care and maintenance: to prevent damage to engineered wood, it should be cleaned regularly and dust and dirt removed from it. Use cleaners made of engineered wood. You can also use wood conditioners to keep moisture inside the wood. Use of protective coating: using protective coating can prevent wrinkling and scratching. You can use a water, soil, and corrosion protection coating to protect engineered wood from water, weather, insects, and scratches. Coating with paint or wood oil: by coating engineered wood with paint or wood oil, you protect it from moisture, wind, and sunlight. People who want to use engineered wood outdoors can use wood paint to cover the wood to prevent corrosion and scratches.

Installing in the right place: installing engineered wood in environments with moderate temperature and humidity is one of the solutions that can help preserve it. Avoid installing engineered wood in areas that are exposed to water, sunlight, and direct wind. In short, to maintain engineered wood, it should be cleaned regularly, use a protective coating, and cover It is important to choose the right place to install it with paint or wood oil [37].

4. Conclusion

Although engineered wood protects the environment and prevents indiscriminate cutting of trees, it should be kept in mind that this type of wood can cause serious harm to human health. Because in its production process, a type of glue is used that contains a harmful substance called formaldehyde; Therefore, and both carpenters and manufacturers are exposed to damage caused by engineered wood; Therefore, these people should look for solutions such as the use of masks and the size of safety equipment to minimize the possible risks of working with engineered wood.

Conflict of interest

"The authors declare no conflict of interest."

Author details

Masuod Bayat
Faculty of Mechanical Engineering, Department of Manufacturing, University of Kashan, Kashan, Iran

*Address all correspondence to: masuod522@gmail.com

References

[1] Schultz TP, Nicholas DD, Preston AF. A brief review of the past, present and future of wood preservation. Pest Management Science: Formerly Pesticide Science. 2007;**63**(8):784-788

[2] Bayat M, Abootorabi MM. Comparison of minimum quantity lubrication and wet milling based on energy consumption modeling. Proceedings of the Institution of Mechanical Engineers, Part E: Journal of Process Mechanical Engineering. 2021;**235**(5):1665-1675

[3] Freeman MH, Shupe TF, Vlosky RP, Barnes HM. Past, present, and future of the wood preservation industry. Forest Products Journal. 2003;**53**(10):8

[4] Preston AF. Can understanding the mechanisms of biodegradation help preservative development?. Washington, DC: ACS Publications; 2003:372-377

[5] Frihart CR. Introduction to special issue: Wood adhesives: Past, present, and future. Forest Products Journal. 2015;**65**(1-2):4-8

[6] Broda M, Hill CA. Conservation of waterlogged wood—Past, present and future perspectives. Forests. 2021;**12**(9):1193

[7] Townsend TG, Solo-Gabriele H, editors. Environmental Impacts of Treated Wood. United States of America: CRC Press; 2006

[8] Olorunnisola AO. The past, present and future outlook of the wood industry in Nigeria. In: Wood Industry-Past, Present and Future Outlook. London, UK: IntechOpen; 2023

[9] Preston AF. Wood preservation. Forest Products Journal. 2000;**50**(9):12

[10] Ding Y, Pang Z, Lan K, Yao Y, Panzarasa G, Xu L, et al. Emerging engineered wood for building applications. Chemical Reviews. 2022;**123**(5):1843-1888

[11] Mao Y, Hu L, Ren ZJ. Engineered wood for a sustainable future. Matter. 2022;**5**(5):1326-1329

[12] Milner HR, Woodard AC. 8 - Sustainability of engineered wood products. In: Jamal MK, editors. Woodhead Publishing Series in Civil and Structural Engineering. Sustainability of Construction Materials (Second Edition). Australia: Woodhead Publishing; 2016. pp. 159-180. ISBN: 9780081009956. DOI: 10.1016/B978-0-08-100370-1.00008-1

[13] Bodig J, Jayne BA. Mechanics of wood and wood composites. bibliotecadigital.infor.cl, Santiago. 1982

[14] Kumar C, Leggate W. An overview of bio-adhesives for engineered wood products. International Journal of Adhesion and Adhesives. 2022;**118**:103187

[15] Walsh-Korb Z, Avérous L. Recent developments in the conservation of materials properties of historical wood. Progress in Materials Science. 2019;**102**:167-221

[16] Plomion C, Leprovost G, Stokes A. Wood formation in trees. Plant Physiology. 2001;**127**(4):1513-1523

[17] Handbook W. Wood as an Engineering Material. Gen. Tech. Rep. FPL–GTR–113. Madison, Wisconsin,

USA: Forest Products Laboratory. USDA Product Society; 1999

[18] Yadav R, Kumar J. Engineered wood products as a sustainable construction material: A review. In: Engineered Wood Products for Construction. London, UK: IntechOpen; 2021

[19] Ellefson PV, Stone RN. US Wood-Based Industry: Industrial Organization and Performance. USA: Praeger Publishers; 1984

[20] Baharin A, Fattah NA, Bakar AA, Ariff ZM. Production of laminated natural fibre board from banana tree wastes. Procedia Chemistry. 2016;**19**:999-1006

[21] Dong Y, Cui X, Yin X, Chen Y, Guo H. Assessment of energy saving potential by replacing conventional materials by cross laminated timber (CLT)—A case study of office buildings in China. Applied Sciences. 2019;**9**(5):858

[22] Kumar D, Mandal A. Review on manufacturing and fundamental aspects of laminated bamboo products for structural applications. Construction and Building Materials. 2022;**348**:128691

[23] Sharma B, van der Vegte A. Engineered bamboo for structural applications. In: Nonconventional and Vernacular Construction Materials. United States: Woodhead Publishing; 2020. pp. 597-623

[24] Merrild H, Christensen TH. Recycling of wood for particle board production: Accounting of greenhouse gases and global warming contributions. Waste Management & Research. United Kingdom. 2009;**27**(8):781-788

[25] Wikipedia contributors. File: Particle board-cross section scan.jpg - Wikipedia [Internet]. Available from: https://en.wikipedia.org/wiki/File:Particle_board-cross_section_scan.jpg

[26] Nishimura T. Chipboard, oriented strand board (OSB) and structural composite lumber. In: Wood Composites. Woodhead Publishing. Elsevier Ltd. 2015. pp. 103-121

[27] Raheem D. Application of plastics and paper as food packaging materials-An overview. Emirates Journal of Food and Agriculture. 2013:**25**(3):177-188. DOI: 10.9755/ejfa.v25i3.11509

[28] Kubba S. Chapter 6 choosing materials and products. In: Kubba S, editor. Green Construction Project Managament and Cost Oversight. USA: Elsevier; 2010. pp. 221-266

[29] Wikipedia contributors. Medium-density fibreboard. Wikipedia. 2023. Available from: https://en.wikipedia.org/wiki/Medium-density_fibreboard

[30] Lee TC, Pu'ad NM, Selimin MA, Manap N, Abdullah HZ, Idris MI. An overview on development of environmental friendly medium density fibreboard. Materials Today: Proceedings. 2020;**29**:52-57

[31] Pugazhenthi N, Anand P. A Review on Mechanical Properties of Medium Density Fiberboard Prepared from Different Fiber Materials. In: Yang LJ, Haq A, Nagarajan L, editors. Proceedings of ICDMC 2019. Lecture Notes in Mechanical Engineering. Singapore: Springer; DOI: 10.1007/978-981-15-3631-1_28

[32] Young, Timothy M., et al. Improving Innovation from Science Using Kernel Tree Methods as a Precursor to Designed Experimentation. Applied Sciences 10.10. Switzerland. 2020:3387

[33] Wikipedia contributors. Hardboard. Wikipedia [Internet]. 26 May, 2023. Available from: https://en.wikipedia.org/wiki/Hardboard

[34] Nasir M, Khali DP, Jawaid M, Tahir PM, Siakeng R, Asim M, et al. Recent development in binderless fiber-board fabrication from agricultural residues: A review. Construction and Building Materials. 2019;**211**:502-516

[35] Acda MN. High-density fiberboard from wood and keratin fibers: Physical and mechanical properties. Current Materials Science: Formerly: Recent Patents on Materials Science. 2022;**15**(2):154-163

[36] Chen C, Kuang Y, Zhu S, Burgert I, Keplinger T, Gong A, et al. Structure–property–function relationships of natural and engineered wood. Nature Reviews Materials. 2020;**5**(9):642-666

[37] Khademibami L, Bobadilha GS. Recent developments studies on wood protection research in academia: A review. Frontiers in Forests and Global Change. 2022;**5**:793177

Chapter 3

Engineered Wood Products from Planted Tropical Timber Species

Razak Wahab, Mohamad Saiful Sulaiman, Johari Zainuddin, Hashim W. Samsi, Sofiyah Mohd Razali, Abdul Fattah Ab Razak and Ellisha Iling

Abstract

Engineered wood products (EWP) have gained popularity and recognition in Malaysia's construction industry. These products refer to a category of wood products that are manufactured by bonding or combining wood strands, veneers, or fibers with adhesives to create a stronger and more stable material compared to solid wood. In Malaysia, the use of EWP, such as plywood, laminated veneer lumber, glued laminated timber, and particleboard, has been growing steadily. These products offer several advantages over traditional solid wood, including improved strength, dimensional stability, and resistance to warping and splitting. EWP is also often used as a sustainable alternative to solid wood because it utilizes smaller, fast-growing trees and reduces waste. EWP find applications in various construction projects, including residential, commercial, and industrial buildings. They are commonly used for interior and exterior structural elements, such as beams, columns, trusses, and flooring systems. EWP, such as plywood and particleboard, are also used extensively for wall and roof sheathing, furniture manufacturing, and decorative applications. The Malaysian construction industry has recognized the benefits of EWP in terms of cost-effectiveness, design flexibility, and environmental sustainability. As a result, there has been increased adoption of these products in both large-scale projects and smaller construction ventures.

Keywords: planted tropical wood species, engineered wood products, economic aspect, designs of engineered wood houses, components for house construction

1. Introduction

Engineered wood made from a variety of wood types, such as recycled woods, hardwoods, and softwoods. In addition to being utilized for furniture and cabinetry, it is frequently employed as a building material for walls, roofs, and floors. The strength and stability of engineered wood, which is less prone to warping, cracking, and shrinking than conventional solid wood, is one of its advantages. Additionally, because it utilizes fewer natural resources and can be made from recycled wood, it has a smaller environmental impact than solid wood [1, 2]. In comparison with conventional solid wood [3, 4] and other building materials like concrete [5], engineered wood products have a number of advantages.

Products made of engineered wood are designed to be more durable and sturdier than those of solid wood. Using adhesives, wood strands, veneers, or fibers are bonded or layered together throughout the manufacturing process to produce a material with consistent strength and dimensional stability. Engineered wood is less likely to twist, warp, or split, resulting in greater long-term performance [6–8].

Some of the key advantages of the engineered wood include that it can be produced in a range of dimensions, forms, and arrangements to satisfy particular design and building specifications. It makes architectural design more flexible and makes it possible to build intricate, cutting-edge structures. Engineered wood products, such as glulam (glue-laminated timber), allow architectural versatility by being built into vast spans and curved shapes [9].

The engineered wood products are often considered more environmentally friendly than solid wood as they make efficient use of timber resources by utilizing smaller, fast-growing trees and incorporating by-products and residues from the wood industry. Engineered wood also reduces waste since it can be manufactured in large panels or beams, minimizing the need for cutting down large trees. Furthermore, the manufacturing process can utilize adhesives with low volatile organic compound (VOC) emissions, contributing to improved indoor air quality [10].

In addition, the engineered wood products go through meticulous manufacturing procedures to guarantee reliability and consistency. Engineered wood products, in contrast to natural solid wood, are designed to satisfy particular performance criteria and may be tested and certified in accordance with those standards. Natural solid wood can vary naturally in strength and features. It is possible to better plan the construction process and design structures thanks to the predictability of material qualities.

Products made of engineered wood are frequently more affordable than those made of solid wood. Smaller trees, which are typically more economical and accessible, can be used to make them. Large panels or beams can also be made from engineered wood, which eliminates the need for intricate joinery or assembly. Over time, engineered wood's greater dimensional stability also results in less waste and cheaper maintenance expenses [11, 12].

Certain types of engineered wood products, such as fire-rated plywood or fire-resistant particleboard, can offer improved fire resistance compared to solid wood. These products are designed to meet specific fire safety regulations and can be used in applications where fire protection is a concern [13, 14].

It is important to note that the benefits of engineered wood products can vary depending on the specific product and its intended application. It is always recommended to consult industry professionals and adhere to relevant standards and guidelines when using engineered wood in construction projects [14].

2. Literature review

The development of engineered wood products in Malaysia dates back to the middle of the twentieth century, when the nation started looking into alternatives to solid wood for use in the building and furniture sectors. Here is a general history of engineered wood products in Malaysia, albeit the precise timeframe and milestones may differ [15, 16].

In Malaysia, plywood, one of the earliest types of engineered wood, became well-liked in the 1950s. Utilizing adhesive, thin veneer layers of wood are bonded together to create plywood. The local construction sector benefited greatly from its importation, which was initially done to accommodate the rising demand for building supplies [17].

In the 1960s and 1970s, particleboard emerged as another engineered wood product in Malaysia. Particleboard is produced by compressing wood particles or chips with resin under heat and pressure. It provided an alternative to solid wood in furniture manufacturing and interior applications.

Laminated veneer lumber (LVL) gained prominence in Malaysia in the 1990s. LVL is made by bonding veneer sheets together with adhesives to create strong and dimensionally stable structural members. LVL found applications in beams, columns, and other load-bearing elements in construction.

Early in the new millennium, glue-laminated timber, or glulam, gained popularity in Malaysia. In order to construct larger, stronger, and more visually beautiful structural elements, glulam involves bonding layers of solid wood together. Curved beams and arches were among the architectural and structural uses for glulam.

The engineered wood product known as cross-laminated timber, or CLT, is relatively new and has attracted interest on a global scale. CLT panels, which are formed by stacking and gluing many layers of wood at right angles, have been utilized in construction projects all over the world because of their strength and sustainability, albeit the specific adoption of CLT in Malaysia may differ.

The adoption of engineered wood products over the years in Malaysia has been driven by factors such as the availability of raw materials, advancements in manufacturing technologies, and the desire for sustainable and cost-effective construction solutions. Malaysian manufacturers have invested in production facilities and research to enhance the quality and range of engineered wood products available in the local market. It is important to note that the specific milestones and advancements in engineered wood products in Malaysia beyond my knowledge cut-off in September 2021 may require more up-to-date sources or industry reports to provide the most accurate and recent information.

3. The economic of engineered wood for construction

Engineered wood as shown in **Figure 1**, also known as composite wood, is a type of wood product made by joining wood fibers, strands, or veneers together with adhesives, resins, or other materials. It is used in many different applications, such as construction, furniture, and flooring. Engineered wood has grown in popularity as a housing building material in the recent years. This is due to a variety of factors, including its strength and durability, low cost compared to traditional wood, and environmental friendliness. [3, 18, 19].

One of the main advantages of engineered wood for housing is its strength and durability. Because it is manufactured using a combination of wood fibers and adhesives, engineered wood is less susceptible to warping, cracking, and splitting than traditional wood. It is also less likely to be affected by moisture, insects, and other environmental factors that can damage wood [20].

Figure 1.
Some of the engineered wood components used in Malaysia.

Figure 2.
Engineered wood components for large and medium construction works.

Another advantage of engineered wood is its cost-effectiveness. While traditional wood can be expensive, especially if it is of high quality or sourced from remote locations, engineered wood can be manufactured using a wide range of materials and techniques (**Figures 1** and **2**), making it more affordable and accessible to builders and homeowners.

Moreover, engineered wood is also eco-friendly because it is made from wood fibers and other materials that are often derived from sustainable sources, and it is a renewable resource that can help to reduce the environmental impact of housing construction. Additionally, it is proved by manufacturing using fewer raw materials and less energy than traditional wood, and it is a more sustainable and environmentally friendly option overall [19].

The economic benefits of using engineered wood for housing are significant. From its strength and durability to its cost-effectiveness and eco-friendliness, it offers a range of advantages that make it an ideal building material for modern homes. As such, it is likely to continue to grow in popularity and become an increasingly important part of the housing industry in the years to come [21].

4. The designs of engineered wood for houses

The design of engineered wooden houses is similar to that of traditional wooden houses, but there are some key differences due to the unique properties of engineered wood [3, 18, 19]. The engineered wood can be fabricated to the various available sizes and shapes. Therefore, it was easier to create customized designs and structures. So, that allows for greater flexibility and creativity in the design process, as well as the ability to optimize the use of materials and space, especially combined with bracket joining components (**Figure 3**) [3].

In another hand, engineered wood is stronger and more durable compared to traditional wood. Apart from that, it can be used to create larger and more complex structures. This means that designers can incorporate features such as curved walls, cantilevered roofs, and expansive windows that would be difficult or impossible to achieve with traditional wood. An additional benefit of using engineered wood in the design of wooden houses is that it allows for more efficient construction because engineered wood is manufactured to precise specifications, it can be pre-cut and pre-drilled off-site, reducing the time and labor required for on-site assembly among

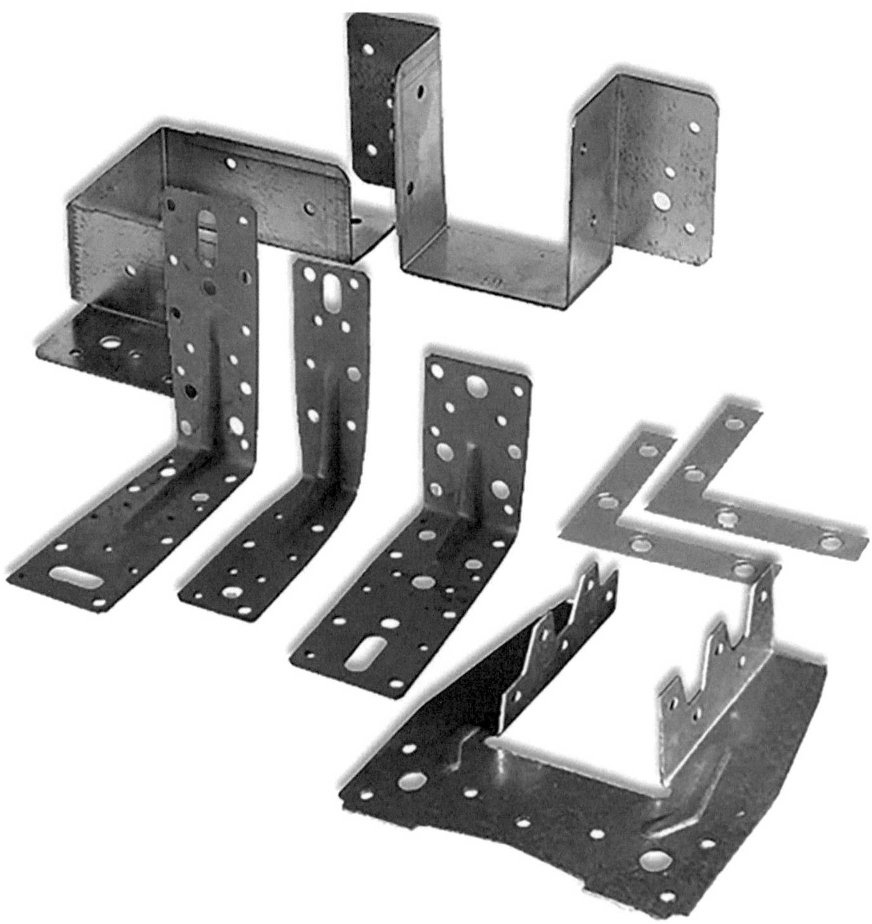

Figure 3.
Metal braces of various sizes and uses to join different components in house construction.

Figure 4.
T-plate joint are commonly used to join the wood beams and poles.

of joining types as represented in **Figures 4–7**. Moreover, engineered wood is also lighter and more consistent in quality than traditional wood, and it is easier to handle and transport, further reducing construction time and costs by applying the connecting joint between floor and wall as highlighted in **Figures 8** and **9**. Finally, the use of engineered wood in the design of wooden houses has environmental benefits. Because engineered wood is made from recycled or sustainably sourced materials and requires less energy to manufacture than traditional wood, it is a more environmentally friendly choice. Additionally, because engineered wood is more durable than traditional wood, it can reduce the need for frequent replacements and repairs, further reducing environmental impact [19].

The design of engineered wooden houses offers a range of benefits over traditional wooden houses, including greater flexibility, increased strength and

Figure 5.
Universal end joint plate.

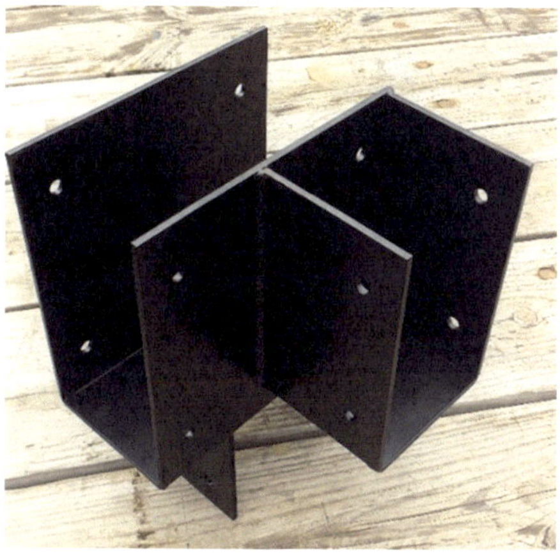

Figure 6.
L-plate joint.

Figure 7.
Plate for four joint.

Figure 8.
Plate to join the floor and the wall.

Figure 9.
Another type of plate to join the floor and the wall.

durability, more efficient construction, and environmental sustainability. As such, it is an ideal choice for modern homes and is likely to continue to grow in popularity in the years to come.

5. House component construction

Laminated or cross-laminated timber houses, also known as wooden houses in the tropic, typically consist of several components that work together to create a strong and durable structure [22–24].

Some of the main components for house construction include:

The foundation of a timber house is typically made of concrete or masonry, and it provides a solid base for the rest of the structure [25].

Wall framing: The walls of a timber house are typically made of wooden studs that are spaced at regular intervals and connected with plates and other framing members.

The roof of a timber house is typically made of wooden rafters that are connected to the walls and support the roof covering.

Sheathing is a layer of wood or other material that is attached to the exterior of the wall and roof framing to provide additional strength and rigidity.

Siding: Siding is the outermost layer of the wall that provides protection against the elements and adds esthetic appeal. Common siding materials include wood, vinyl, and fiber cement.

The roof covering protects the house from the weather and includes materials such as shingles, metal, or tile.

Windows and doors: These are openings in the walls that provide access and light. Wooden houses typically use wooden windows and doors to match the overall esthetic.

Insulation is used to improve the energy efficiency of the house and keep it comfortable year-round. Common insulation materials include fiberglass, cellulose, and spray foam.

Figure 10.
Another design of the universal end joint braces.

The interior finishes that include the materials such as drywall, trim, and flooring that are used to finish the inside of the house and create a comfortable living space.

Wall framing is an essential component of a wood house. It involves creating a structure that will support the weight of the roof and the other elements of the house [26–28].

All of these components are joint together in creating a strong and stable. Metallic braces shown in **Figures 10** and **11** are quite commonly used.

Some of the steps involved in wall framing for a wood house are as follows:

The first step is to create a plan for the wall framing. This will involve determining the size and spacing of the studs, as well as the placement of windows and doors.

Once the plan is in place, the wall can be laid out on the building site. This will involve measuring and marking the locations of the studs and other framing members.

Next, the wood for the wall framing can be cut to size. This will typically involve using a saw to cut the studs, plates, and other framing members to the desired length.

With the lumber cut to size, the wall can be assembled. This will involve nailing or screwing the studs and other framing members together according to the plan.

Once the wall is assembled, it can be raised into place. This will typically require the help of several people to lift the wall and position it correctly.

Once the wall is in place, it will need to be secured to the foundation and the adjacent walls. This will involve using nails, screws, or bolts to attach the wall to the

Figure 11.
A universal corner braces.

Figure 12.
The T-metallic braces used between the laminated timber beam and the pole.

other elements of the house. **Figures 12–14** show the metallic braces used in joining laminated beams with the timber poles.

Once the wall is secured, a layer of sheathing can be added to provide additional strength and rigidity. This may involve using plywood, OSB, or another material to cover the exterior of the wall framing.

Finally, windows and doors can be installed in the wall framing. This will involve cutting openings in the sheathing and framing to accommodate the windows and doors, and then installing them according to the manufacturer's instructions.

Roof framing is another critical component of a timber house that supports the roof's weight and provides structural stability [29–31].

The steps are involved in roof framing:

Before starting the roof framing process, you need to design the roof, which will involve choosing the roof style, calculating the pitch or slope, and determining the load-bearing capacity.

The ridge beam is the horizontal beam that runs along the roof's peak, and it is the roof's highest point. The ridge beam is supported by the walls and should be the first component of the roof framing installed.

Figure 13.
Metallic brackets are used to strengthened the structure between the house beams and poles.

Figure 14.
The metallic braces or brackets are used to strengthen the roof structure.

The rafters are the sloping beams that connect the ridge beam to the wall plates. They need to be installed at regular intervals, and their size and spacing will depend on the roof design and load-bearing capacity.

Collar ties or ridge boards are horizontal members installed between opposing rafters to provide additional structural stability.

Purlins are horizontal beams that are installed perpendicular to the rafters and support the roof covering. They may be necessary for certain roof types or heavy roofing materials like slate or tile.

Roof sheathing is a layer of material installed over the rafters and purlins, providing a flat surface for the roof covering to be installed. Plywood, OSB, or other materials can be used for sheathing.

The final step in roof framing is to install the roof covering. The covering can be made of a variety of materials, including shingles, metal, tile, or slate.

Roof framing requires careful planning, accurate measurements, and knowledge of structural engineering. It is essential to follow building codes and regulations to ensure that the roof framing is safe and meets all standards.

Sheathing is an important component of a timber house that is installed on the exterior walls and roof framing. Its primary purpose is to provide a strong, flat surface for the installation of exterior finishes and to add structural rigidity to the building [32].

Some common sheathing materials used in wood house construction include the following:

Plywood is a popular choice for sheathing because it is strong, durable, and resistant to moisture. It is typically made of thin layers of wood veneer that are glued together with the grain direction alternating between layers.

Oriented strand board (OSB) is a type of engineered wood product that is made from wood strands that are bonded together with resin and wax. It is less expensive than plywood and provides good structural strength [33].

Fiberboard is a type of wood-based panel that is made from wood fibers that are bonded together with a resin binder. It is lightweight and easy to cut, and provides good insulation value.

Gypsum board, also known as drywall, is a type of sheathing that is made of gypsum plaster sandwiched between two layers of paper. It is commonly used as an interior sheathing material but can also be used on the exterior in certain applications.

Insulating sheathing is a type of sheathing that provides both structural support and insulation value. It is typically made of foam plastic insulation with a rigid facing material like plywood or OSB.

When selecting a sheathing material for a wood house, it is essential to consider the climate, the intended use, and the local building codes and regulations. Proper installation of sheathing is also critical to ensure that it functions correctly and provides adequate structural support.

Siding is an important component of a wood house as it provides protection from the elements, enhances curb appeal, and adds to the house's overall esthetic [34–36].

Some common types of siding for a wood house are as follows:

Wood siding is a traditional and popular choice for a wood house. It is available in a variety of species, including cedar, pine, and redwood. Wood siding can be painted or stained, and it is easy to replace individual boards if necessary.

Vinyl siding is a low-maintenance and affordable option for a wood house. It is available in a variety of colors and styles, including those that mimic the look of wood siding. Vinyl siding can be easily cleaned with soap and water and does not require painting.

Fiber cement siding is a durable and low-maintenance option for a wood house. It is made of cement, sand, and cellulose fibers, and it is available in a variety of colors and styles, including those that mimic the look of wood siding.

Brick or stone veneer: Brick or stone veneer can add a timeless and classic look to a wood house. It is available in a variety of styles and colors, and it is durable and of low maintenance.

Stucco is a popular choice for a wood house with a southwestern or Mediterranean style. It is made of a mixture of cement, sand, and lime, and it can be painted in a variety of colors.

When selecting a siding material for a wood house, it is essential to consider the climate, the intended use, and the local building codes and regulations. Proper installation of siding is also critical to ensure that it functions correctly and provides adequate protection from the elements.

Roof covering is the material that is installed over the roof framing to provide protection from the elements [37, 38].

Some common types of roof coverings for a wooden house are as follows:

Asphalt shingles are the most common type of roof covering for a wooden house. They are affordable, easy to install, and available in a wide range of colors and styles. They typically last 15–30 years depending on the quality. Metal roofing is a durable and long-lasting option for a wooden house. It is available in a variety of materials, including steel, aluminum, and copper, and it can last up to 50 years or more. Metal roofing is resistant to fire, insects, and rot, and it is energy-efficient.

Wood shingles or shakes are a traditional and classic choice for a wooden house. They are made from cedar or redwood, and they can last up to 30 years or more. They require regular maintenance, including cleaning, sealing, and occasional replacement of damaged or worn shingles.

Clay or concrete tiles are a popular choice for a wooden house with a Spanish-style architecture. They are durable and long lasting, with a lifespan of up to 50 years or more. They are available in a variety of colors and styles.

Slate is a high-end and long-lasting option for a wooden house. It is made from natural stone, and it can last up to 100 years or more. It is available in a variety of colors and styles, and it is highly resistant to fire and insects.

When selecting a roof covering for a wooden house, it is essential to consider the climate, the intended use, and the local building codes and regulations. Proper installation of the roof covering is also critical to ensure that it functions correctly and provides adequate protection from the elements.

Windows and doors are essential components of a wooden house. They provide natural light, ventilation, and access to the outdoors.

Some common types of windows and doors for a wooden house are as follows [39–42]:

Single-hung windows are a classic and traditional choice for a wooden house. They consist of a fixed upper sash and a lower sash that slides up and down to allow ventilation.

Double-hung windows are similar to single-hung windows but with two movable sashes that slide up and down. They are a more modern and convenient option for a wooden house.

Casement windows are hinged at the side and swing outward like a door. They provide excellent ventilation and are a good choice for areas with a lot of wind.

Sliding windows consist of two or more sashes that slide horizontally past each other. They are a popular choice for modern and contemporary wooden houses.

French doors are a classic and elegant choice for the entrance to a wooden house or a patio. They consist of two doors that open outward from the center and provide a wide opening for easy access.

Sliding doors consist of two or more panels that slide horizontally past each other. They are a popular choice for a wooden house with a modern and minimalist style.

When selecting windows and doors for a wooden house, it is essential to consider the style, energy efficiency, and local building codes and regulations. Proper installation of windows and doors is also critical to ensure that they function correctly and provide adequate insulation and security.

Insulation is a critical component of a wooden house as it helps to reduce energy costs, improve indoor comfort, and protect against moisture and air infiltration.

Some common types of insulation for a wooden house are as follows:

Fiberglass insulation is a popular and affordable option for a wooden house. It is made of glass fibers and is available in batts or blown-in form. Fiberglass insulation has a high R-value (a measure of its resistance to heat flow) and can be used in walls, ceilings, and floors.

Cellulose insulation is made of recycled paper and is available in blown-in form. It has a high R-value and is a good choice for a wooden house with high ceilings or irregularly shaped spaces.

Spray foam insulation is a high-performance and energy-efficient option for a wooden house. It is available in two types: open-cell and closed-cell. Open-cell foam is less dense and is used for interior walls and ceilings, while closed-cell foam is denser and is used for exterior walls and roofs.

Mineral wool insulation is made of rock or slag fibers and is available in batts or blown-in form. It has a high R-value and is a good choice for a wooden house in areas with high winds or extreme temperatures.

Rigid foam insulation is a board-like material that is available in a variety of types, including polystyrene, polyurethane, and polyisocyanurate. It has a high R-value and is a good choice for a wooden house with a basement or crawl space.

When selecting insulation for a wooden house, it is essential to consider the climate, the local building codes and regulations, and the type of construction. Proper installation of insulation is also critical to ensure that it functions correctly and provides adequate protection against moisture and air infiltration.

Interior finishes for a wooden house play an essential role in creating a comfortable and inviting living space.

Some common types of interior finishes for a wooden house are as follows [43, 44]:

Drywall is a popular and affordable option for the walls and ceilings of a wooden house. It is made of gypsum plaster sandwiched between two layers of paper and is available in different sizes and thicknesses.

Wood paneling is a classic and traditional option for the walls and ceilings of a wooden house. It can be made of various types of wood, including pine, cedar, and oak, and is available in different styles, such as tongue-and-groove or shiplap.

Paint is a versatile and affordable option for the walls and ceilings of a wooden house. It is available in a wide range of colors and finishes and can be used to create different moods and styles.

Stucco is a durable and low-maintenance option for the interior walls of a wooden house. It is made of cement, sand, and lime and is applied in several layers to create a smooth or textured finish.

Tile is a popular and stylish option for the floors, walls, and backsplashes of a wooden house. It is available in different materials, such as ceramic, porcelain, and natural stone, and can be used to create different patterns and designs.

Carpet is a comfortable and cozy option for the floors of a wooden house. It is available in different colors and textures and can be used to add warmth and softness to a room.

When selecting interior finishes for a wooden house, it is essential to consider the style, durability, and maintenance requirements. Proper installation of interior finishes is also critical to ensure that they function correctly and provide a comfortable and inviting living space.

6. Engineered wood for tropical houses

There is a growing body of research on the use of engineered wood in housing construction, including wooden engineered houses [4, 20, 45].

Here are some key data points:

The cost of building a wooden engineered house can vary widely depending on factors such as the size and complexity of the design, the type of engineered wood used, and the cost of labor in the area. However, in general, wooden engineered houses are often less expensive to build than traditional wooden houses because they require less raw material and can be manufactured more efficiently.

Wooden engineered houses are often more durable than traditional wooden houses because engineered wood is less susceptible to warping, cracking, and splitting. Additionally, engineered wood is often treated with preservatives or other materials that can help to prevent rot, decay, and insect damage.

Wooden engineered houses can be designed to be highly energy efficient by incorporating features such as insulation, air sealing, and efficient heating and cooling systems. This can help to reduce energy consumption and lower heating and cooling costs over time.

The use of engineered wood in housing construction can have significant sustainability benefits because it is often made from recycled or sustainably sourced materials and requires less energy to manufacture than traditional wood. Additionally, because engineered wood is more durable than traditional wood, it can reduce the need for frequent replacements and repairs, further reducing the environmental impact.

Wood engineered houses are generally considered to be safe and structurally sound, provided they are designed and built to code. However, as with any type of building material, it is important to ensure that the construction is carried out by qualified professionals and that the house is maintained properly over time.

Figures 15–21 show some of the ready timber structures made up with either the laminated or cross-laminated wood for house construction.

The data on wooden engineered houses suggest that they offer a range of benefits over traditional wooden houses, including lower cost, greater durability, improved energy efficiency, environmental sustainability, and safety. As such, they are an ideal choice for modern homes and are likely to continue to grow in popularity in the years to come.

Figure 15.
The frame and floor structure made from the glue-laminated.

Figure 16.
The beams and poles of a house made from Acacia mangium laminated timber.

Figure 17.
The authors standing in front of a glue-laminated bridge made from mixed tropical timber species.

Figure 18.
The close-up view of the joinery between the bridge and the foundation using metal.

Figure 19.
A cross-laminated structure of a cross-laminated house of size 38 sq. m.

Figure 20.
The front view of the cross-laminated model house.

Figure 21.
A timber glue-laminated gallery structure in Johore Bahru, Malaysia.

7. Recommendations

If you are considering building an engineered wooden house, then you need to keep in mind some of these recommendations:

Building an engineered wooden house requires specialized knowledge and expertise. Look for a builder who has experience with engineered wood construction and a track record of successful projects. There are many different types of engineered wood available, each with its own unique properties and characteristics. Consult with your builder to determine which type of engineered wood is best suited to your needs and budget.

Engineered wooden houses can be designed to be highly energy efficient, which can help to reduce your heating and cooling costs over time. Consider incorporating features such as insulation, air sealing, and efficient heating and cooling systems into your design. While engineered wood is generally more durable than traditional wood, it still requires regular maintenance to ensure that it stays in good condition over time. Make sure to plan for regular inspections and maintenance as part of your overall building plan. As with any type of building material, it is important to ensure that your engineered wooden house is designed and built to code to ensure safety and structural integrity. Work with your builder to ensure that all safety requirements are met.

Building an engineered wooden house can be a great choice for those looking for a durable, energy-efficient, and environmentally friendly housing option. With careful planning and attention to detail, an engineered wooden house can provide you with a beautiful and sustainable home for many years to come. Engineered wood structure with a good design and process properly can last long (**Figure 21**).

8. Conclusions

The benefits of using the tropical engineered wood products include strength and durability, sustainability, design flexibility, cost-effectiveness, consistent performance, and quality. These benefits make the engineered wood products a valuable material for various applications in construction and other industries. In order word, the economic benefits of using engineered wood for housing are significant. From its strength and durability to its cost-effectiveness and environmental friendliness, it offers a range of advantages that make it an ideal building material for modern homes. Therefore, it is likely to continue to grow in popularity and become an increasingly important part of the housing industry in the coming years. Furthermore, engineered wood house designs offer many benefits over traditional wooden house, including greater flexibility, increased strength and durability, more efficient construction, and environmental sustainability. It offers an ideal choice for modern house and is likely to continue to grow in popularity in the years to come. In another part, roof framing calls for meticulous design, precise measurements, and structural engineering expertise. To guarantee that the roof frame is secure and complies with all requirements, it is crucial to follow construction laws and regulation.

Nevertheless, when choosing insulation for a wooden house, it is important to consider the climate, local building codes and regulations, and the type of construction. Proper insulation installation is also important to ensure it functions properly and provides adequate protection against moisture and air infiltration. In terms of choosing interior finishes for a wooden house, it is important to consider style, durability, and maintenance requirements. The correct installation of interior finishes is also important to ensure that they function properly and provide a comfortable and attractive living space. Overall, the data on engineered wood houses show that they offer a range of benefits over traditional wooden houses, including lower costs, better durability, improved energy efficiency, environmental sustainability, and safety. It is therefore an ideal choice for modern homes and is likely to continue to grow in popularity in the years to come. Building an engineered wooden house can be the best option for those looking for a durable, energy efficient and environmentally friendly housing option. With careful planning and attention to detail, an engineered wood house can provide you with a beautiful and sustainable home for years to come.

Acknowledgements

We the authors would like to express our appreciation for the financial support provided by the University of Technology Sarawak (UTS) through its UTS Flagship Research Grants 2022.

Author details

Razak Wahab[1*], Mohamad Saiful Sulaiman[1], Johari Zainuddin[2], Hashim W. Samsi[3], Sofiyah Mohd Razali[1], Abdul Fattah Ab Razak[1] and Ellisha Iling[1]

1 Centre of Excellence in Wood Engineered Products, University of Technology Sarawak (UTS), Sibu, Malaysia

2 Sarawak Timber Industry Development Corporation, Kuching, Malaysia

3 Forest Research Institute, Kuala Lumpur, Malaysia

*Address all correspondence to: drrazakw5181@uts.edu.my

References

[1] Sulaiman MS, Razali SM, Kalahari Journals, Wahab R, Edin T, Mokhtar N, et al. Evaluation of thermal treatment on the physical and mechanical properties of *Acacia Hybrid*. International Journal of Mechanical Engineering. 2022;**7**(4):218-226

[2] Wahab R, Sulaiman MS, Samsi HW, Ghani RSM, Mokhtar N. The Effectiveness of Oil-Heat Treatment in the Main Chemical Constitutes for Planted 15-Year-Old Cultivated *Acacia Hybrid*. Asian Journal of Science and Technology. 2020;**10**(12). ISSN: 0976 3376

[3] Yadav R, Kumar J. Engineered Wood Products as a Sustainable Construction Material: A Review. Engineered Wood Products Construction; IntechOpen. 2021. DOI: 10.5772/intechopen.99597

[4] Milner HR, Woodard AC. Sustainability of engineered wood products. In: Sustainability of Construction Materials. Woodhead Publishing Series in Civil and Structural Engineering; 2016. pp. 159-180

[5] Ahmed S, Arocho I. Analysis of cost comparison and effects of change orders during construction: Study of a mass timber and a concrete building project. Journal of Building Engineering. 2021;**33**:101856

[6] Sulaiman MS, Razali SM, Wahab R, Edin T, Mokhtar N, Ab Razak AF, et al. Evaluation of thermal treatment on the physical and mechanical properties of *Acacia Hybrid*. International Journal of Mechanical Engineering (7). Special Issue 4, 2022. pp. 218-226. Kalahari Journals

[7] Wahab R, Edin T, Mokhtar N, Sulaiman MS, Ghani RSM and Razak MH. Monitoring Changes in the Colour, Strength and Chemical Properties of Oil Heat Treated 18-Years Old Cultivated *Acacia mangium*. Chapter 5 in a Book "Recent Research Advances in Biology. Book Publisher International: 2020; Vol. 4". DOI: 10.9734/bpi/rrab/v4. ISBN: 978-93-90516-91-9. eBook ISBN: 978-93-90516-92-6

[8] Wahab R, Mazalan INSA, Samsi HW, Sulaiman MS, Ghani RSM, Mokhtar N. Performance in Accelerated Laboratory Tests of Oil Heat Treated 16-Year-Old *Acacia mangium*. BOOK Chapter in Prime Archives in Agricultural Research. 18 pages. Vide Leaf Publication. 2020. ISBN: 978-81-944664-0-6

[9] Izyan K, Wahab R, Mahmud S, Othman S, Affendy H, Hanim RA, et al. International Journal of Chemistry. 2010;**2**(1):97-107. ISSN: 1916-9701 Canadian Center of Science and Education

[10] Wahab R, Ghani RSM, Sulaiman MS, Edin T, Mokhtar N, Razak MH. Improvement in Durability of Oil Heat Treated 16-Year-Old *Acacia mangium* in Laboratory Tests. Chapter 7 in a Book Cutting-Edge Research in Agricultural Sciences. Book Publisher International. 2020. Vol. 5. pp. 92-106. DOI: 10.9734/bpi/cras/v5. ISBN: 13 (15) 978-93-90516-76-6

[11] Mokhtar N, Edin T, Wahab R, Ghani SM, Sulaiman MH, Razak MH, et al. Properties of the Oil Heat-Treated 10 & 15 years-old cultivated *Tectona grandis*. A Book Chapter in New Visions in Science and Technology. 2021. Vol. 1. pp. 64-84. Book Publisher International. ISBN-13 (15) 978-93-90516-76-6

[12] Sulaiman MS, Wahab R, Mokhtar N, Edin T, Razali SM, Ghani RSM. Scanning Electron Microscopy Study of the Effectiveness Oil Heat Treatment on 10-years old teak wood in ground contact test. Borneo Journal of Sciences and Technology. 2021;**3**(2):24-32. e-ISSN: 2672-7439

[13] Wahab R, Mohd Fauzi N, Mokhtar N, Sulaiman MS, Mohd Ghani RS, Edin T. Enhancing Mechanical Properties Of *Rhizophora Apiculata* Through Engineered Laminated Boards. The Journal Agriculture and Forestry Časopis Poljoprivreda i šumarstvo. 2020e;**66**(3):53-64. ISSN 1800-9492 (Online). DOI: 10.17707/Agricult Forest

[14] Khalid I, Sulaiman O, Hashim R, Wahab R, Jumhuri N, Rasat MSM. Evaluation on layering effects and adhesive rates of laminated compressed composite panels made from oil palm (Elaeis guineensis) fronds. Journal of Materials and Design. 2015;**68**:24-28

[15] Jumaat MZ, Rahim AHA, Othman J, Razali FM. Timber engineering research and education in Malaysia. UM Research Repository. 2006. pp. 2494-2497

[16] Ghani RSM, Wahab R, Mustafa NMC, Mokhtar N, Sulaiman MS, Lee M. Physical and Mechanical Properties of Cassava-Bamboo Composite Lumber. Journal of Engineering and Science Research. 2018;**2**(6):06-09

[17] Ghani RSM, Wahab R, Azmi SNB, Wi KM, Mokhtar N, Sulaiman MS. Comparison of Properties between Solid and Laminated Mahang Wood. Comparison of Properties between Solid and Laminated Mahang Wood Advanced Journal of Technical and Vocational Education. 2019;**2**(2):24-28

[18] Ilgın HE, Karjalainen MP. Perceptions, Attitudes, and Interests of Architects in the Use of Engineered Wood Products for Construction: A Review. Engineered Wood Products for Construction: 2021:83-95

[19] Manninen H. Long-Term Outlook for Engineered Wood Products in Europe. European Forest Institute; Technical Report 91, 2014

[20] Kutnik M, Suttie E, Brischke C. Durability, efficacy and performance of bio-based construction materials: Standardisation background and systems of evaluation and authorisation for the European market. In Performance of bio-based building materials. Elsevier: Woodhead Publishing. 2017:593-610

[21] Rasat MSM, Wahab R, Yunus AAM, Moktar J, Ramle SFM, Kari ZA et al. Physical and mechanical properties of bio-composite board from compressed oil palm fronds. Advances in Natural and Applied Sciences. 2014;7(5):572-582

[22] Benson TH. The Timber-Frame Home: Design, Construction, Finishing, A Fine homebuilding book. Taunton Press; 1988

[23] Jeff DL, editor. American Institute of Timber Construction (AITC). Timber Construction Manual. John Wiley & Sons; 2012. ISBN: 978-0-470-54509-6

[24] Sobon JA, Schroeder R. All About Post-and-Beam Building. Timber Frame Construction. Storey Publishing, LLC; 1984. p. 208

[25] Wahab R, Samsi HW, Sulaiman MS, Ghani RSM, Mokhtar N. Properties of Altered Matured *Bambusa Vulgaris* via Heat Treatment Process, Research Journal of Pharmaceutical Biological and Chemical Sciences. 2019;**10**(4):257-274

[26] Satheeskumar N, Henderson DJ, Ginger JD, Wang C. Finite element

modelling of the structural response of roof to wall framing connections in timber-framed houses. Engineering Structures. 2017;**2017**(134):25-36

[27] Tussenbroek GV. Timber-framed town houses in the northern Netherlands before 1600: Construction and geographical distribution. Vernacular Architecture. 2017;**48**:44-62

[28] Satheeskumar N. Wind load sharing and vertical load transfer from roof to wall in a timber-framed house. Doctoral dissertation. Australia: James Cook University; 2016

[29] Seliutina L, Ratkova E, Okulova E. Wooden roof structures of the Moberg house. In E3S Web of Conferences. EDP Sciences; 2023:**376**

[30] Dutkiewicz M, Hajyalikhani P, Lamparski T, Whitman L, Covarrubias J. Structural bracing of wooden roofs under the extreme winds. IOP Conference Series: Materials Science and Engineering. 2021;**1203**:022024. DOI: 10.1088/1757-899X/1203/2/022024

[31] Vilčeková S, Harčárová K, Moňoková A, Burdová EK. Life Cycle Assessment and indoor environmental quality of wooden family houses. Sustainability. 2020;**12**(24):10557

[32] Pynkyawati T, Lesmana E, Sukendar D. Utilization of used wood as material structure and construction of Sitinggil house buildings. International Journal of Built Environment and Scientific Research. 2021;**5**(2): 97-108

[33] Pásztory Z, Peralta PN, Molnár S, Peszlen IM. Modeling the hygrothermal performance of selected north American and comparable European wood-frame house walls. Energy and Buildings. 2012;**49**:142-147

[34] Jim BA, McFarland H, Carli M, Jacobs G, Harry P. Environmental Assessment of House Cladding Products. Dovetail Partners Consuming Responsibly Report No. 11. 2019

[35] Patterson J. Vinyl foam: Effect of density on physical properties. Journal of Vinyl and Additive Technology. 1998:**4**(1):26-29

[36] Damery DT, Fisette P. Decision making in the purchase of siding: A survey of architects, contractors, and homeowners in the US northeast. Forest Products Journal. 2001;**51**(7/8):29-36

[37] Cîmpeanu A. Measuring roof coverings of vernacular architecture in open-air museums. A long-term approach for sustainable intervention work culture. Society Economy Politics. 2023;**2**:100-110. DOI: 10.2478/csep-2022-0013

[38] Ayşe BS, Elçin T. Single score environmental performances of roof coverings. Sustainability. 2023;**2023**(15):4387. DOI: 10.3390/su15054387

[39] Wan-Geon L. A study on the windows and doors of traditional houses in Jecheon. Korean Institute of Interior Design Journal. 2013;**22**(4):94-103

[40] Hui Y. Decorative arts on manor house wood doors and windows of Changjia. Key Engineering Materials. 2011;**480-481**:1289-1292. DOI: 10.4028/www. scientific.net/KEM.480-481.1289

[41] Ahn E, Kim J. Windows and doors lattice structure for improving energy efficiency. Wireless Personal Communications. 2014;**79**:2415-2423. DOI: 10.1007/s11277-014-1766-3

[42] Priadi T, Nandika D, Sofyan K, Acmad Witarto AB. Biodeteriorasi

komponen kayu rumah di beberapa daerah yang berbeda suhu dan kelembabannya. Jurnal Ilmu dan Teknilogi Hasil Hutan. 2010;**3**(1):26-31

[43] Abdul Halim N. Interior floor finishes preferences for Malaysian houses. Doctoral dissertation. Malaysia: Universiti Teknologi MARA; 2022

[44] Saito Y, Nishimaki M, Tatsumi Y, Ono S, Kinoshita M, Sasayama S, et al. Effects of wooden and vinyl interior finishes on stress reduction estimate by biological and psychological parameters. Mokuzai Gakkaishi/Journal of the Japan Wood Research Society. 2009;**55**(2):101-107

[45] Ayanleye S, Udele K, Nasir V, Zhang X, Militz H. Durability and protection of mass timber structures: A review. Journal of Building Engineering. 2022;**46**:103731

Chapter 4

Standard Test Methods for Elastic and Shear Properties

Ahmed Mohamed

Abstract

A fundamental requirement for efficient use of timber-based composite structures like glulam beams is an accurate knowledge of their mechanical behavior and the material properties characterizing that behavior. Determining the elastic properties, such as the modulus of elasticity and shear modulus, for glulam beams requires careful experimentation and can be challengeable due to the anisotropic nature of wood. Determining these properties is not as simple and straightforward as in isotropic materials. Shear tests, such as torsion and shear field and compression loading tests, are commonly employed to determine the shear modulus of glulam beams. To determine the modulus of elasticity, experimental methods such as bending and compression tests are commonly used. In this chapter, we will discuss the experimental methods commonly used to determine the modulus of elasticity and shear modulus, for timber-based composite structures. These properties are crucial for understanding the structural behavior and design of these materials. This chapter describes the commonly used methods, bending tests, torsion tests, and compression loading tests, in determining their values. To obtain accurate and reliable results, it is essential to conduct these experimental methods following established standards and carefully controlling the test conditions, specimen preparation, loading configurations, and measurement techniques.

Keywords: modulus of elasticity, shear modulus, timer-based composite, glulam beams, experimental methods

1. Introduction

The mechanical properties of wood composites rely on different parameters, including wood species, the type of adhesives used to glue the wood pieces, and the geometry of the wood pieces [1]. Mechanical properties play a vital role in evaluating the characteristics of wood-based composites. The elastic and strength properties provide important information for material selection, design, and establishing product specifications [1]. Elastic properties relate the resistance of a material to deformation under an applied load and the ability of the material to recover its original dimensions when the load is removed [1]. Known as elastic constants, the elastic properties include the basic and fundamental properties such as shear modulus (G) and modulus of elasticity (MOE).

The shear modulus (G), known as modulus of rigidity, is a material constant which relates shear stress to shear strain and indicates the resistance to deformation caused by shear stresses [1]. It is a fundamental mechanical property of wood that is used in the design of timber and engineered wood products. The shear modulus is critical when designing for lateral torsional buckling of the timber beams [2]. G is also significant in designing serviceability of wood-joist floors [3] and is an important input for setting up analytical and finite element models [4]. The modulus of elasticity (MOE), also known as the elastic modulus or Young's modulus, quantifies how much a material deforms under an applied stress within the elastic range. MOE can be calculated from the slope of the linear portion of the stress-strain graph obtained during a tensile or compressive test.

The European standard CEN, and the American Standard of Testing Methods (ASTM) have developed several test methods for the determination of shear and elastic properties of timber. These methods provide guidelines for testing timber in small clear wood specimens as well as in structural sizes under flexural and torsion loadings. The European standard CEN [5] provides test methods in laboratory for determining some physical and mechanical properties of in structural sizes timber samples. This standard specifically provides testing guidelines to determine the characteristic values of mechanical properties, including shear modulus and modulus of elasticity, of structural timber and glued laminated timber. The procedures of the test methods recommended by [5] are shown in the sections below.

2. Testing methods for shear modulus

The EN408 standard has proposed the torsion and shear field test methods to determine shear modulus of timber in structural sizes.

2.1 Torsion test method

The torsion test method is a mechanical testing technique used to evaluate shear modulus (G) of materials, including timber. The proposed version of CEN [5] standards provide guidelines for conducting the torsion test on timber. This method involves subjecting a test specimen to a torque along its longitudinal axis, which is achieved by inducing a relative rotation between the supports where the specimen is clamped. The test specimen is clamped at the supports, that are positioned at a distance more than 16 times the cross-sectional depth of the specimen. To measure the response of the specimen, both torque and relative rotational displacements are measured at two specific sections, labeled as Section 1 and Section 2 in **Figure 1**. These sections are located within the free testing length, denoted as l_1. The distance between the supports and these cross-sections, denoted as l_2, should be two to three times the depth of the test specimen. The centers of the supports are aligned in a straight line. This alignment ensures that clamping the test specimen does not cause any deformation that could influence the torsion results. The torque required for the torsional loading can be applied by rotating one or both supports.

The following equation is used to calculate shear modulus (G_{Tor})

$$G_{Tor} = \frac{K_{Tor} l_1}{\eta\, h\, b^3} \tag{1}$$

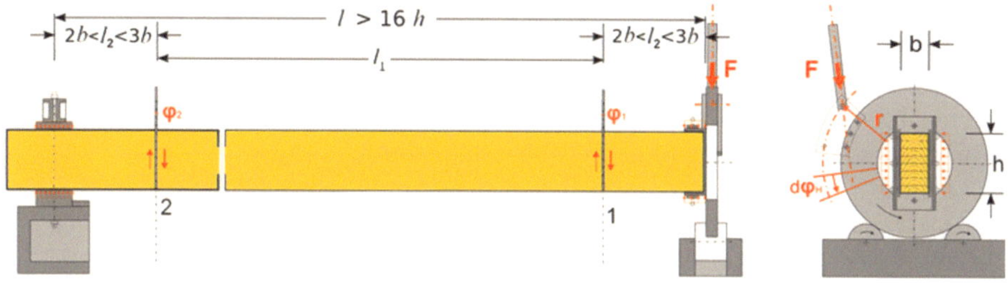

Figure 1.
Example of test setup with requirements of specific gauge locations [5].

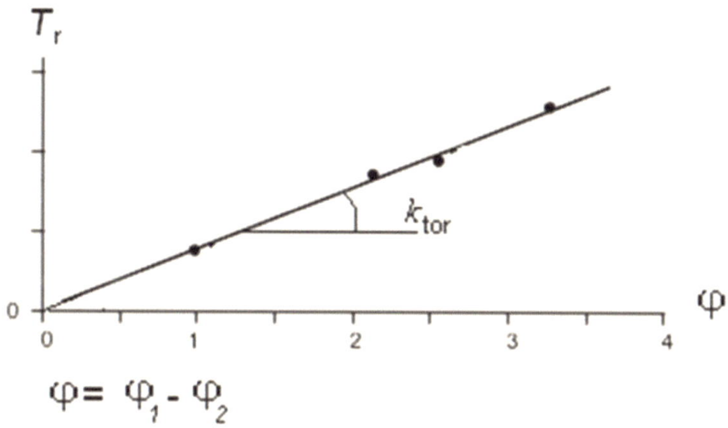

Figure 2.
Torque versus relative rotation [5].

h/*b*	1	1.2	1.5	2	2.5	3	4	5	10
η	0.1406	0.166	0.196	0.229	0.249	0.263	0.281	0.291	0.312
χ	0.4158	0.4564	0.4618	0.4904	0.5162	0.5334	0.5634	0.5960	0.6270

Table 1.
Shape factor values for torsion test [5].

The torque stiffness, K_{Tor}, can be determined using a linear regression analysis as shown in **Figure 2**. Linear regression analysis can be conducted on the linear elastic portion of the graph of torque and relative twist of a specimen within its proportional limits. Where:

l_1 is the gauge length, h is the depth of the specimen, b is the width of the specimen,

η is the shape factor that is dependent on the depth to width ratio of the specimen and can be obtained from **Table 1**.

2.2 Shear filed test method

The shear field test method is a recommended approach for determining the shear modulus of structural-size timber beams. It involves conducting standardized four-point bending tests on the timber beams according to the CEN [5] standard. The objective of this test is to measure the shear distortion within the areas of constant transverse force. Precise measuring instruments are placed on both sides of the constant transverse force, opposite each side of the beam. These instruments accurately measure length displacements across the shear fields of the sample during testing.

The setup of the shear field test is shown in **Figure 3**. According to the test method, the test specimen should be symmetrically loaded in bending at two points over a span of 18 times its cross-sectionals' depth. However, if these requirements cannot be met exactly, the distance between the load points and the supports can be changed by a maximum of 1.5 times the specimen depth. The span and length of the test specimen can also be changed by a maximum of three times its depth while maintaining the symmetry of the test. To minimize local indentation, small steel plates are recommended to be inserted between the specimen and the loading heads or supports. The length of these plates should not exceed half the depth of the specimen.

The applied load during the test should be applied at a constant rate. In the middle of the area under constant shear stress, a square is marked on both side faces of the specimen. These squares are placed symmetrically with respect to the height of the test piece. A device that measures the change in the diagonals of the square is fixed to the test specimen at the square corners, as shown in **Figure 4**. The shear deformation

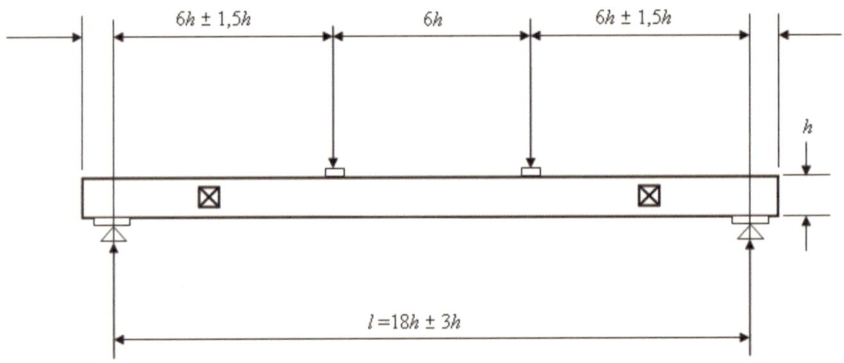

Figure 3.
Test SETUP for shear field test [5].

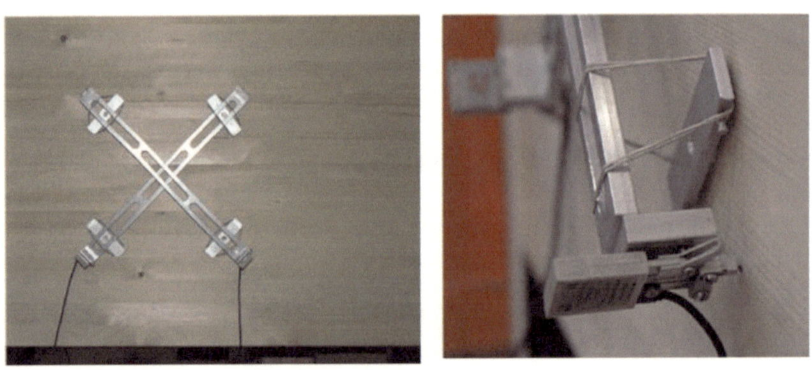

Figure 4.
Example of the shear field test apparatus fixed on one of both sides [5].

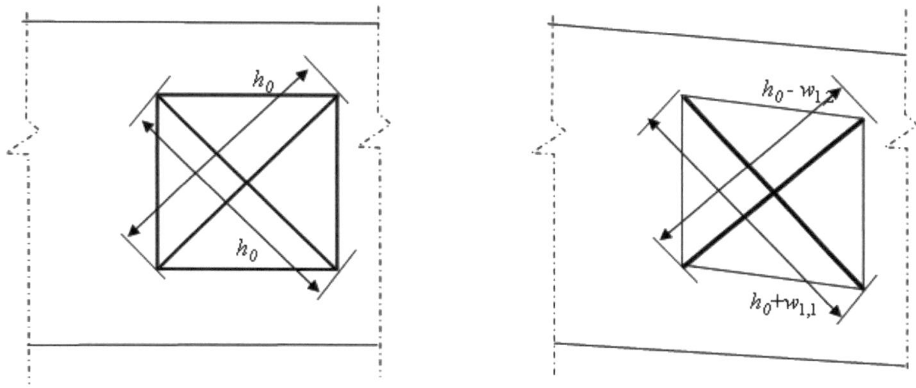

Figure 5.
Deformation of the square with diagonals [5].

can be determined from these measurements by calculating the mean value of the summation of the absolute readings of both diagonals at each side face of the cross-section, as shown in **Figure 5**. This measurement provides the necessary data to calculate the shear modulus of the timber beam using the shear field test method.

The shear modulus $G_{tor,s}$ is given by the equation:

$$G_{tor,s} = \alpha \frac{h_0}{bh} \frac{(V_{s,2} - V_{s,1})}{(w_2 - w_1)} \tag{2}$$

Where:

$$\alpha = \frac{3}{2} - \frac{h_0^2}{4h^2} \tag{3}$$

$$w_i = \frac{(|w_{i,1}| + |w_{i,2}|)}{2} with i = 1, 2 \tag{4}$$

w_i is the mean deformation of both diagonals on opposite side faces of the beam for a given shear load $V_{s,i}$, in millimeters.

$V_{s,2} - V_{s,1}$ represents the shear load increment, which is the difference between the shear load values, in Newton.

3. Test methods for determining the modulus of elasticity (MOE)

There are static and dynamic test methods that can be used to determine the MOE of materials. The static methods are used more commonly and are based on the relationship between the load and the deformation of the sample, while the dynamic methods are based on the relationship between the frequency of vibration of a sample and its MOE.

The European standard CEN [5] provides various test methods to obtain the modulus of elasticity values of structural timber and glued laminated timber. These methods are used to measure the modulus of elasticity in different conditions such as bending, tension parallel and perpendicular to the grain, and compression parallel and perpendicular to the grain. The following sections provides review of some of these test methods.

3.1 Four-point bending test method

The CEN edition [6] introduced a global bending modulus of elasticity and renamed the existing test as the local MOE. The four-point bending test method can be used to determine both the local and global MOE [5]. In the four-point bending test, a full-size specimen is loaded symmetrically in bending at two points over a span that is 18 times the depth of the section as provided in **Figure 6**. If it is not possible to meet this exact span requirement, the distance between the load points and the supports can be changed by up to 1.5 times the specimen depth. Additionally, the span and length of the test specimen can be changed by up to three times its depth, as long as the symmetry of the test is maintained. To minimize local indentation, small steel plates can be inserted between the specimen and the loading heads or supports. The length of these plates should not exceed one-half of the depth of the specimen. The test method also suggests that the test specimen should be supported simply, and lateral restraint should be provided to prevent lateral torsional buckling. This allows the specimen to deflect without significant frictional resistance. The applied load during the test should be at a constant rate, as recommended by the standard. The test setup for determining the local MOE in bending is provided in **Figure 6**.

To determine the local modulus of elasticity (MOE) in bending, the deformation is measured at the center of a central gauge length, which is defined as 5 times the depth of the section. The deformation and load are observed at two different times during the test, denoted as (w_1, F_1) and (w_2, F_2) in **Figure 7**. It is crucial to ensure that these two measurements are made within the proportionality limit of the beam. The recommended approach is to take the deformation as the average of measurements on both side faces at the neutral axis.

The equation to calculate the local modulus of elasticity, $E_{m,l}$ of structural size specimen in the four-point bending test is

$$E_{m,l} = \frac{a l_1^2 (F_2 - F_1)}{16 I (w_2 - w_1)} \tag{5}$$

Where:

$E_{m,l}$: Local modulus of elasticity in the four-point bending test, in Newtons per square millimeter (N/mm^2).

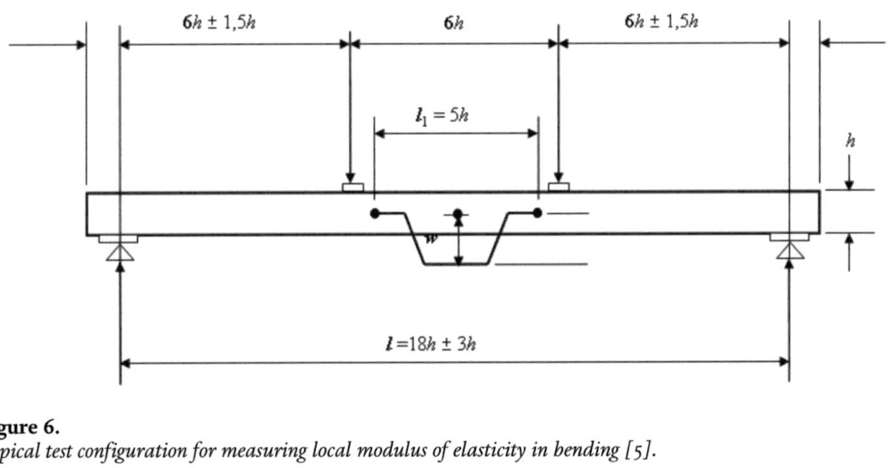

Figure 6.
Typical test configuration for measuring local modulus of elasticity in bending [5].

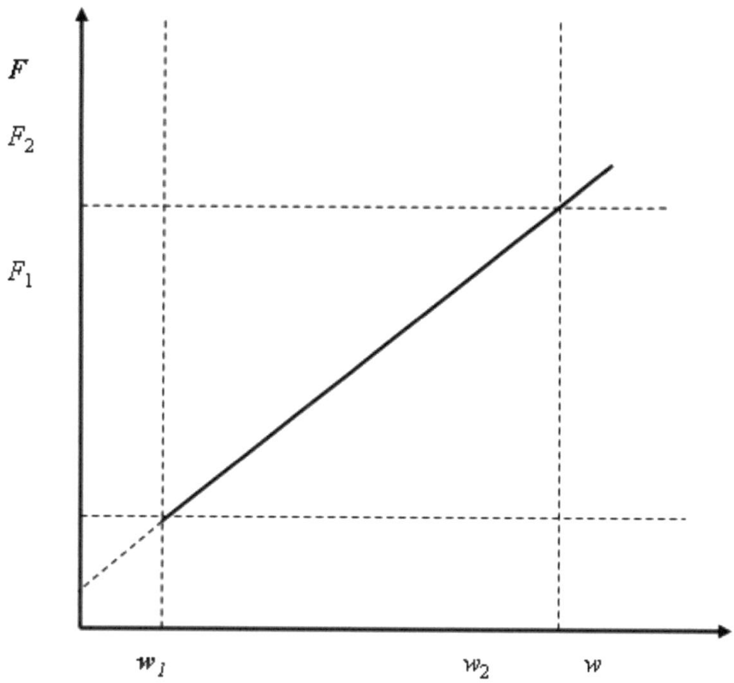

Figure 7.
Load-deformation graph within the elastic range of deformation [5].

F_1 and F_2: Loads observed at two different times during the test, measured in Newton (N),

w_1 and w_2: Average deformations measured on both side faces at the neutral axis at the corresponding times, measured in millimeters (mm).

I: Moment of inertia of the beam's cross-sectional shape, in (mm^4).

l_1: Gauge length for the determination of modulus of elasticity, measured in millimeters (mm).

a: Distance between a loading position and the nearest support, in millimeters (mm).

The global MOE in bending is evaluated using the same test configuration as the one used for measuring the local MOE. The test arrangement is illustrated in **Figure 8**.

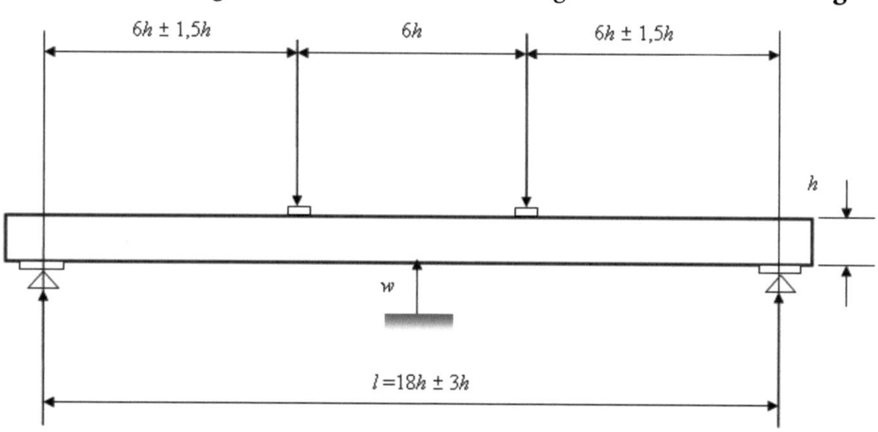

Figure 8.
Typical test configuration for measuring the global MOE in bending [5].

To determine the global MOE, the deformation is measured at two locations:

1. At the center of the span.

2. From the center of the tension or compression edge.

Additionally, when the deformation is measured at the neutral axis, it is recommended to take the average of measurements on both side faces of the test specimen.

The equation to calculate the global modulus of elasticity, $E_{m,g}$ of structural size specimen in the four-point bending test is.

$$E_{m,g} = \frac{3al^2 - 4a^3}{2bh^3\left(2\frac{w_2 - w_1}{F_2 - F_1)} - \frac{6a}{5Gbh}\right)} \tag{6}$$

Where:

l: The span in bending, measured in millimeters (mm).

b: The width or the smaller dimension of the cross-section, also measured in millimeters (mm).

h: The depth or the larger dimension of the cross-section, measured in millimeters (mm).

G: The shear modulus, which is taken as $G = 650$ N/mm^2.

3.2 Compression test methods

The compression test is one of the most used methods for determining the modulus of elasticity and other elastic and strength properties of materials. The specific characteristics of the compression tests for measuring MOE parallel to grain and MOE perpendicular to grain in structural and glued-laminated timber, as outlined in the CEN [5] standard.

1. MOE Parallel to Grain Test:

   This test is widely used for determining the modulus of elasticity of timber parallel to the grain direction. The deformation, or strain, of the specimen is measured over a central gauge length.

2. MOE Perpendicular to Grain Test:

   The perpendicular to grain test is specifically conducted to assess the modulus of elasticity of timber when compressed perpendicular to the grain. In this test, the MOE is reported at the proportional limit.

3.2.1 Test specimen

According to CEN [5] standard, the test specimen requirements for the parallel to grain compression tests are as follows:

The test specimen should have a full cross-section.

The length of the specimen should be six times the smaller cross-sectional dimension.

The two contact surfaces of the specimen must be accurately prepared to ensure they are parallel to each other and aligned perpendicular to the longitudinal axis of the specimen.

Table 2 provides the dimensions of the test specimen for structural and glued laminated timber samples in compression perpendicular to grain. These samples are shown in **Figures 9** and **10**.

3.2.2 Conditioning test specimen

The conditioning of the test specimens for structural and glued-laminated timber should be carried out in a controlled environment with specific temperature and relative humidity conditions. The test specimens should be conditioned at a temperature of 20 ± 2 degrees Celsius (°C). The relative humidity during conditioning should be set at 65 ± 5%. The specimens should be conditioned until they reach a state of constant mass. This is achieved when the results of two successive weightings do not differ more than 0.1% of the mass of the test specimen.

Specimen characteristics						
Structural timber			Glued laminated timber			
b(mm)	h(mm)	l(mm)	Volume (m^3)	$b \times 1$ (mm^2)	b minimum (mm)	h (mm)
Tension						
45	180	70	0.01	25,000	100	400
Compression						
45	90	70	—	25,000	100	200

Where:
b: *The width of the cross-section or the smaller dimension of the cross-section. Measure in mm.*
h: *The height of the test specimen perpendicular to the grain, measured in mm.*
l: *The length of the test piece between the testing machine grips in compression and tension, measured in mm.*

Table 2.
Dimension of structural timber or glued laminated timber test pieces [5].

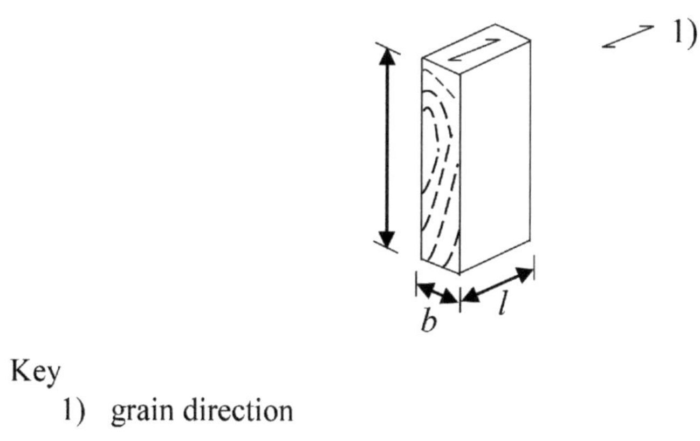

Figure 9.
A typical structural timber test specimen [5].

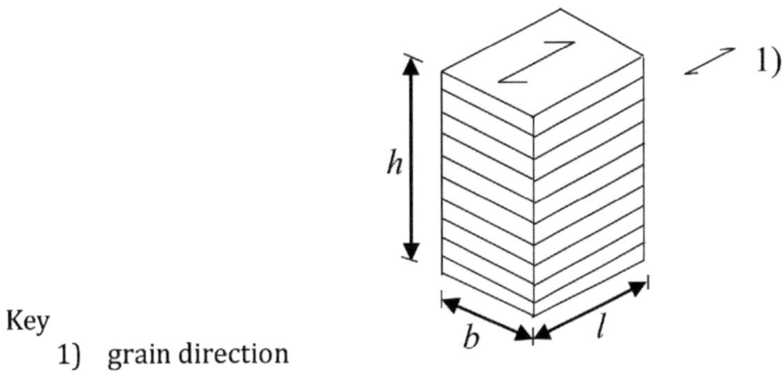

Figure 10.
A typical glued laminated timber test specimen [5].

3.2.3 Loading procedure

The CEN (European Committee for Standardization) test configuration for measuring the modulus of elasticity (MOE) in compression is outlined in **Figure 11**. The test specimen needs to be positioned vertically between the platens of the testing machine. The test specimen needs to be centrally loaded to ensure that the compressive force is applied uniformly. This is achieved by using spherically seated loading-heads. These loading-heads allow the application of compressive loads without causing any bending or uneven loading. After pre-loading the test specimen, the spherical loading head should be securely fastened to prevent slipping. The test specimen should be loaded uniformly with a constant rate of loading throughout the test. The applied load during the test should be measured with an accuracy of 1%.

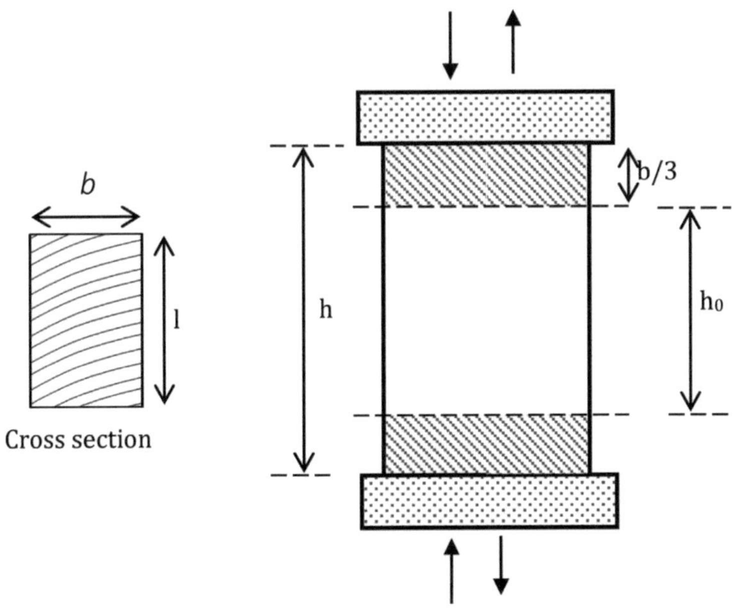

Figure 11.
Typical compression test configuration in accordance with [5].

To minimize the effects of distortion during the MOE compression test, the use of displacement sensors, such as extensometers, is recommended. Two displacement sensors (extensometers) should be positioned symmetrically in the middle of the two opposite surfaces of the test sample. The readings from the two displacement sensors should be averaged to obtain a more reliable and representative result. The displacement sensors should be positioned at least at a distance from the contact surfaces of one third of the largest cross-sectional size of the test sample. This distance helps to negate the grip effect, which refers to the influence of the grips or clamps used to hold the specimen during testing.

In the CEN test configuration, the deformation in compression parallel to the grain is measured over a central gauge length. This gauge length should be four times the smaller cross-sectional dimension of the test specimen. The deformation in compression perpendicular to the grain is measured over a gauge length denoted as h_0. The gauge length h_0 is approximately 0.6 times the height of the test specimen, h. The gauge length h_0 must be located centrally in the height of the test specimen. It should be positioned at a distance of at least b/3 from the loaded ends of the test specimen. Here, b represents the width or thickness of the test specimen, depending on its orientation.

3.2.4 Calculation of MOE in compression parallel to the grain

The equation for calculating the modulus of elasticity parallel to the grain in compression $E_{c,0}$ according to [5] standard, is as follows:

$$E_{c,0} = \frac{l_1(F_2 - F_1)}{A(W_2 - W_1)} \tag{7}$$

Where:

l_1 refers to the span in bending, measured in (mm),

$F_2 - F_1$ represents the increment of load on the straight-line portion of the load-deformation curve. (Measured in Newton),

$W_2 - W_1$ indicates the increment of deformation corresponding to $F_2 - F_1$ (measured in millimeters),

A is the initial crossed section of the sample.

3.2.5 Calculation of MOE in compression perpendicular to the grain

The modulus of elasticity perpendicular to the grain in compression is assessed with the same test configuration for measuring MOE parallel to the grain. According to EN 408 standard, $E_{c,90}$ may be determined using the iterative process as follows:

- Estimate the maximum compression load $E_{c,90,\,\max}$.
- Using the test results, plot a graph that represents the relationship between the applied load and the resulting deformation as sown in **Figure 12**.
- Calculate two points on the load-deformation curve corresponding to 0.1 times and 0.4 times the estimated maximum compression load. These points represent 0.1 ($E_{c,90,\,\max}$) and ($E_{c,90,\,\max}$) respectively.

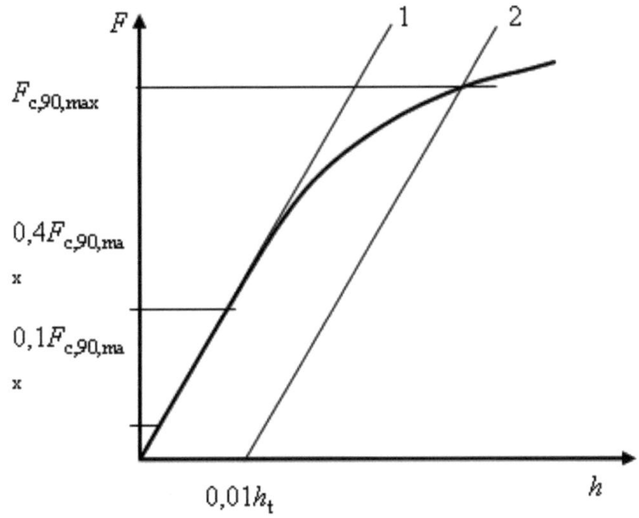

Figure 12.
Load-deformation graph for compression perpendicular to the grain according to [5].

- Find the intersection point of the load-deformation curve with a straight line (line 1) drawn between the points representing 00.1 ($E_{c,90,\,\max}$) and ($E_{c,90,\,\max}$).
- This intersection point represents the deformation corresponding to ($E_{c,90,\,\max}$) determine the intersection of these two points in the load-deformation curve. Through these two values draw a straight line 1 as shown in **Figure 12**
- Draw a straight line (line 2) parallel to line 1. Line 2 originates from the load axis (F = 0) and is located at a distance equivalent to a deformation of 0.01 h_0, where $h_0$represents the original thickness of the specimen.
- Find the point of intersection between line 2 and the load-deformation curve. The deformation value at this intersection point represents the modulus of elasticity perpendicular to the grain in compression $E_{c,90,\,\max}$.

The equation for calculating the modulus of elasticity perpendicular to the grain in compression, $E_{c,90}$, according to the EN 408:2012 standard, is as follows:

$$E_{c,90} = \frac{(F_{40} - F_{10})h_0}{(w_{40} - w_{10})\,bl} \tag{8}$$

Where:
F_{40}: Load at 0.4 of $E_{c,90,\,\max}$ (N).
F_{10}: Load at 0.1 of $E_{c,90,\,\max}$ (N).
w_{40}: Deformations at F_{40} (mm).
w_{10}: Deformations at F_{10} (mm).
h_0: Gauge length (mm).
b: Width of the specimens (mm).
l: Length of the specimens (mm).

Additional information

This chapter is adapted from my PhD thesis: Mohamed, A. S. Photogrammetric and Stereo Vision Techniques for Evaluating Material Properties in Timber and Timber-Based Composite Structures. (Thesis). Edinburgh Napier University. Retrieved from http://researchrepository.napier.ac.uk/Output/462281

Author details

Ahmed Mohamed
Edinburgh Napier University, Edinburgh, UK

*Address all correspondence to: a.mohamed2@napier.ac.uk

References

[1] Laboratory FP and UDo Agriculture. The Encyclopedia of Wood. United States: Skyhorse Publishing Inc.; 2007

[2] EN 1995-1-1. Eurocode 5-Design of Timber Structures-Part 1-1: General-Common Rules and Rules for Buildings. Brussels, Belgium: European Committee for Standardization; 2004

[3] Foschi RO. Structural analysis of wood floor systems. Journal of the Structural Division. 1982;**108**(7): 1557-1574

[4] Chui Y. Application of ribbed-plate theory of predict vibrational serviceability of timber floor systems. In: The Proceedings of 7th World Conference on Timber Engineering WCTE. Shah Alam, Malaysia; 2002. 4. pp. 87-93

[5] EN 408:2012. Timber Structures-Structural Timber and Glued Laminated Timber- Determination of some Physical and Mechanical Properties Perpendicular to the Grain. Brussels, Belgium: European Committee for Standardization; 2012

[6] EN 408:2003. Timber Structures-Structural Timber and Glued Laminated Timber- Determination of some Physical and Mechanical Properties Perpendicular to the Grain. Brussels, Belgium: European Committee for Standardization; 2003

Chapter 5

Wood Degradation by Fungi and Environmentally Benign Wood Preservatives

Yan Xia and Lu Jia

Abstract

The causes, processes, features and conditional factors of wood decay and degradation due to fungi were reviewed. The degradation path of wood varies depending on diverse fungi and wood species, as fungi possess selectivity in the degradation of versatile wood components. The chemical treatments and preservatives are reviewed to understand their correlation with the decay mechanisms of wood. Environmentally benign wood preservatives are discussed, e.g. one based on chicken feather protein combined with copper and boron salts to replace the traditional wood preservatives together with several environmentally friendly preservatives based on wood extractives as a source of natural raw materials. Excellent functionalities of the protein-based wood preservative suggested that this eco-formulation could offer great potential to be used as an environmentally benign wood preservatives with a more competitive cost. This new system of wood preservatives provides a theoretical basis for further research and the reasonable utilization and scientific protection of wood products.

Keywords: wood degradation, mechanism, fungi, environmentally benign wood preservatives, chemical treatment

1. Introduction

Wood, being a natural resource, exhibits its great value and importance, and can be used widely as a structural material and industrial raw material in many fields of the global economy. It is important to note that, despite wood is a remarkable material, it is susceptible to be biodegraded by the action of microorganisms, such as fungi and bacteria. Thus, wood is ready to be decomposed under proper or certain conditions and returns natural components to ecosystem cycling.

Wood is composed of three main components, namely cellulose, hemicellulose and lignin. The highest content of the three main components is cellulose, a long linear homopolymer, which is the main chemical component in wood cell walls, accounting for approx. 40–50% of the dry weight of wood substrates, composed of β-D-glucose molecules connected by [1–4] glycosidic linkages, and can be broken down into reducing sugars by cellulase. Hemicellulose is also a kind of polysaccharide molecules similar to cellulose, but a heterogeneous material which consists of various monosaccharides, and as such easier to be degraded by microorganisms than cellulose. Lignin, an aromatic heteropolymer, is the third major component of wood, accounting for approx. 25–30% of the wood dry weight [1, 2]. Unlike cellulose and hemicellulose, lignin is a polymer composed of condensed phenylpropane units (including benzene ring and aliphatic structure) with an extremely complex structure and chemical properties. The main function of lignin is to provide intensity and durability, as well as resist attacks from microorganisms and insects. Due to its complex structure, lignin is difficult to be degraded, which is often seen as a difficult obstacle in the production and utilization of wood, and only a few microorganisms, such as white rot fungi, possess the ability to completely degrade lignin [3].

In nature, the decay and degradation of wood products is a complex process that involves the combined action of various microorganisms, which can produce many types of enzymes and disintegrate woody materials as an organic substrate by the secreted enzymes [4]. By the loss of the constituents of wood cell walls, such as cellulose, hemicellulose and lignin, wood will lose its valuable strength and stability, and ultimately be bio-deteriorated by microorganisms. Among the microorganisms, white rot fungi, brown rot fungi, soft rot fungi, and bacteria are the most common causes that contribute to the decay of wood [2, 4]. White rot fungi are a few microorganisms that can completely degrade lignin. Brown rot fungi mainly decompose carbohydrates (cellulose and hemicellulose). Soft rot fungi are always ready to degrade polysaccharides. In addition, there are other microorganisms, such as bacteria, which can also degrade the components of wood [5].

Therefore, the relationship between the chemical components of wood and its degradation mechanism is strong. The degree of wood decay, or the degradation of wood chemical components, and the wood decay mechanisms vary depending on different environmental conditions [2]. For example, humid environments are more prone to cause wood degradation since microorganisms are more ready to colonize and reproduce in humid environments. In addition, different types of wood can also affect diverse types of degradation, leading to different levels of decay.

In order to extend wood service life, some treatments can be performed, such as heat treatment, chemical preservation treatment or other methods [6]. Meanwhile, the in-depth research on the wood degradation mechanism could also provide a theoretical basis and technical support for relieving and preventing wood decay and is also meaningful for both proper protection and reasonable utilization of wood.

2. Wood degradation by fungi

2.1 Wood decay fungi species

Wood decay can mainly be caused by the infection of wood decay fungi. Basidiomycetes [4, 7], which are the most common wood decay fungi in wood and play predominant roles in terrestrial carbon recycling, are well known as members of Basidiomycota, one of two large divisions of *Eumycota*, together with the Ascomycota [8, 9]. Therefore, most wood decay fungi can be assigned to Basidiomycetes or Ascomycetes. In a terrestrial ecosystem, a large group of decomposers, wood rot fungi, has been found, with about 1500 species in Finland and about 2000 species reported in China [10].

In Division Basidiomycota, Gloeophyllales, an order of Class Agaricomycetes, is capable of producing brown rot of wood and contains several important species, among which *Gloeophyllum trabeum* and *Gloeophyllum sepiarium* are two important wood brown rot fungi species in common service above ground [9]. In particular, *G. trabeum* is well known as an important fungus for the test during the decay resistance trial, which has the tolerance of some kinds of organic wood preservatives.

Polyporales is another order of Class Agaricomycetes in Division Basidiomycota and includes many of the fungi species. In Polyporales, *Rhodonia placenta* is another most common species of brown rot fungi and a common test fungus applied in evaluating new wood preservatives. It is interesting that *Trametes versicolor* is another species in Polyporales, but a white rot fungus, which is also a test fungus commonly used for the assessment of the function of wood preservatives especially in hardwoods.

2.2 Inhabiting conditions of decay fungi

Wood decay fungi have certain requirements in inhabiting conditions for growth and survival [11], several factors affecting fungal colonization and propagation are as follows:

1. Water: free water or adequate moisture in wood cell lumina;

2. Oxygen: as living organisms, oxygen is a necessary condition for most wood-inhabiting fungi;

3. Favorable temperature: suitable range of survival temperature for most wood decay fungi is from 15 to 40°C;

4. Available nutritional supplies: sustainable nutrient source is also a necessary factor for long-term living in order to provide sufficient energy and metabolites for synthesis via metabolism;

5. pH range: favorable chemical conditions are necessary for fungal growth on wood, the range, the optimal pH condition, is from 3 to 6.

In-depth comprehension of the fungal inhabiting or living conditions is extremely significant as this information can provide us with a better understanding of how these wood decay fungi survive and multiply, and furthermore how to efficiently protect wood and prevent serious degradation. For example, reducing the surrounding humidity or keeping wood dry below the fiber saturation point (25–30%) could efficiently eliminate the fungi growth. In addition, scientific treatments could be carried out to better protect wood resources, and a new wood preservative system could be developed. Besides, novel biotechnology can also be developed to pretreat the woody raw materials or waste wood even including some organic waste, which can promote the development and progress of environmental protection.

2.3 Wood decay process and feature

2.3.1 Wood decay process

As mentioned before, wood rot fungi are traditionally divided into white rot and brown rot fungi, and different fungi can decompose the different wood chemical compositions due to their own selectivity and action mechanism [12, 13]. During the process of wood degradation, white rot fungi can mainly secrete a key group of extracellular oxidases (oxidative enzymes) to degrade lignin, i.e. lignin peroxidase, manganese (II)-dependent peroxidase, and laccase, which is the most typical and common oxidative enzymes possessing the relative strong degrading ability. Hydrolytic enzymes, such as cellulase, hemicellulase, amylase and pectinase can as well as be secreted by white rot fungi, in consequence, white-rot fungi can completely deconstruct the lignocellulose cell wall materials [4]. In addition, according to the degradation and removal of wood chemical components by white rot fungi, it can be classified into "selective" and "simultaneous" decay path [14]: (1) Selective rot initially degrade the hemicellulose and lignin, but retaining the cellulose [15]; (2) Simultaneous rot degraded cellulose, hemicellulose and lignin in a rather uniform depletion [16, 17]. It is noteworthy that the same white rot fungi can cause selective or simultaneous rot when it decayed different substrates, and even both types rot in the same substrates [18, 19].

Unlike the white-rot fungi, brown rot fungi can secrete a large amount of carbohydrate enzymes, such as cellulase and hemicellulase, pectinase including amylase. Brown rot fungi extensively depolymerize the carbohydrates (cellulose and hemicellulose), leaving the fragments of the degraded cellulose and hemicellulose but retaining the modified lignin which is not depolymerized seriously [20–22]. It has long been thought that these basidiomycetes do not decompose the lignin seriously, and their activities on lignin, the abundant aromatic biopolymer, are limited to minor oxidative modifications [23, 24].

As mentioned before, different enzymes display rather various effects on wood chemical components, especially from various fungi as well as different degradation stages, which could be attributed to the fungi's instinctive motivation and selectivity. Thus, different fungi biodegrade wood in their own selective path, and different biodegradation paths vary between different wood species (soft and hard wood).

2.3.2 Decay features

According to the shape, wood decay can be classified into white rot, brown rot, and soft rot [4]. Most wood decay fungi species are subordinate to Basidiomycota (Basidiomycetes), typically classified into two types, either white- or brown-rot fungi [7]. Brown rot fungi cause significant degradation of cellulose and hemicellulose but with little degradation of lignin, which can only be modified. The typical features of wood brown rot are shrinks and fragmentations, which easily to be decomposed into soft cubic shapes with brown discoloration, due to the lack of cellulose and hemicellulose, and the oxidation of lignin. Conversely, white rot fungi mainly degrade lignin, causing a whitish, needlelike texture or fibrous shape of the decayed wood [25]. Soft-rot fungi, broadly as "non-*Basidiomycete*" destroyers, resemble the brown-rot fungi which utilize exoglucanases, and endoglucanases to degrade cellulose, which was reported that the attack is limited to the amorphous cellulose zones in the microfibrils [26]. Generally, the attack of soft rot fungi is limited primarily to the carbohydrates in cell walls, and limited modification of lignin, such as demethoxylation.

2.4 Variation of wood property

Once wood is infected by fungi, wood degradation will be presented outside or/ and inside of wood, with the result of alterations in chemical compositions, physical

DOI: http://dx.doi.org/10.5772/intechopen.112033

and mechanical properties, and changes of microstructures. Some alterations of wood properties, such as mass losses, chemical components and microstructures are summarized as below:

1. **Mass alteration**

   Wood mass loss is commonly accompanied by most fungal decay due to that fungi need nutrients as they grow through the wood by utilizing the various components of the wood cell wall. Thereby, most wood decay fungi can utilize the various chemical components of the wood substrate in the cell wall as they colonize and grow through the wood, leading to the reduction of the overall wood mass [27]. Wood mass losses can achieve 70% through brown rot, even exceed 95% after white rot and about 5–60% by soft rot. Mass loss will be generally expressed at the different trend of change due to the various decay conditions, which depends on the diverse wood species including the types of decay fungi as well as the different periods during wood decay.

   Mass losses of some wood species in different periods with both brown rot fungi and white rot fungi are presented in **Figure 1** (derived from the research of previous research [28] and authors' results [29]). As portrayed in **Figure 1**, the mass loss of wood clearly increased as the decay time prolongs. Besides, regarding softwood, brown rot fungi were stronger than white rot fungi. In contrast, in hardwood samples, the deconstruction capacity of white rot fungi to hardwood was more aggressive and vigorous than that of brown rot fungi, and this could be that the lignin types of hardwood mainly included guaiacyl lignin and syringyl lignin, and hardwood lignin contained more methoxyl groups, which was more easily decomposed [14, 30]. Furthermore, it was worth mentioning that, in hardwood group, *Hevea brasiliensis* showed more resistance against white rot fungi than *Populus yunnanensis* and *Liquidambar styraciflua*, which may be due to the density of *H. brasiliensis* higher than that of *P. yunnanensis* and *L. styraciflua*, attributing to the significant effect of density on wood properties (**Table 1**).

2. **Changes in chemical components**

   The chemical compositions of diverse wood species at different periods by brown rot and white rot fungi are summarized in **Table 2**.

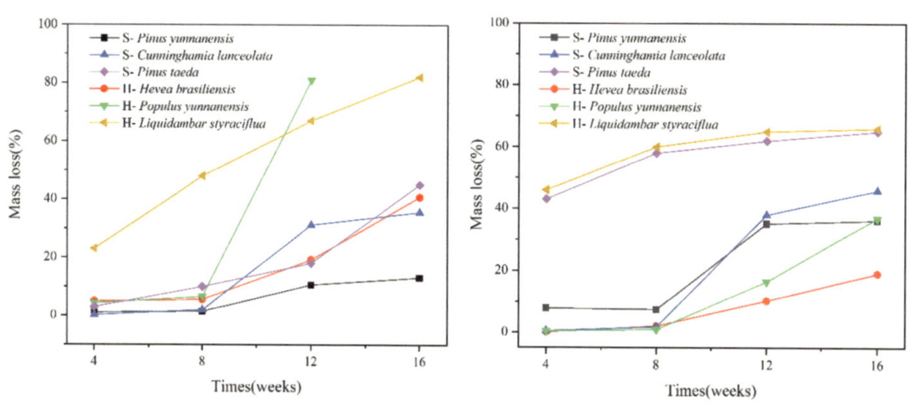

Figure 1.
Mass loss of different wood species decayed by white and brown rot fungi.

Wood species	*P. yunnanensis*	*Cunninghamia lanceolata*	*Pinus taeda*	*H. brasiliensis*	*P. yunnanensis*	*L. styraciflua*
Densities/g. cm^{-3}	0.472	0.401	0.580	0.650	0.364	0.545

Wood densities of different tree species [31–34].

Table 1.
Densities of different wood species.

P. yunnanensis [19]				
Fungi	Time(d)	Glucan (%)	Xylan (%)	Lignin (%)
—	0	33.67	9.73	32.24
T. versicolor	30	25.93	11.32	29.90
	60	26.10	4.49	27.64
	90	33.17	7.18	25.90
G. trabeum	30	30.09	7.72	28.33
	60	25.76	4.36	27.62
	90	23.50	2.97	29.62
R.placenta	30	21.80	7.89	27.05
	60	20.12	6.57	26.89
	90	19.50	3.16	27.95
P. jezoensis [35]				
Fungi	Time(d)	Glucan (%)	Xylan (%)	Lignin (%)
—	0	46.9	23.2	26.8
P. pini	30	45.9	22.2	26.4
	60	40.4	24.2	24.0
	90	31.2	26.3	19.4

Table 2.
Relative chemical compositions of control and degraded wood.

For *P. yunnanensis* wood, *Trametes versicolor* (white rot fungi) caused simultaneous rot, resulting in a rather uniform depletion of glucan, xylan and lignin. Conversely, in brown rot groups, fungi preferentially decomposed carbohydrates, which led lignin retained selectively. However, they decomposed cellulose and hemicellulose in their own selective approach. The xylan decreased from 9.73 to 2.97%, which decayed by *G. trabeum*, while the glucan decreased from 33.67 to 19.50% after *Rhodonia placenta* biodegradation, revealing that *G. trabeum* could degrade hemicellulose selectively, while *Rhodonia placenta* could preferentially attack cellulose. It is interesting that different fungi exhibited obviously different degradation in softwood species, e.g. the *Picea jezoensis* decayed by *Porodaedalea pini*, both cellulose and lignin were depolymerized due to the respective decreased contents, whereas, the increased content of hemicellulose showed the slight degradation of hemicellulose probably due to the weak influence on hemicellulose from this fungus. This also reveals that wood rot fungi depolymerize the chemical compositions of wood substrate through their own degradation pathway.

3. Changes in microstructure

The microstructure of wood will also undergo significant changes after being infected by decay fungi, such as the enlarged porosity in the wood due to the partially or entirely destroyed fibers, which could improve the permeability, decrease the density, and reduce the strength and toughness of wood as well, making it prone to fracture [2]. Some morphology observations are illustrated in **Figure 2** (derived from the authors' results [29]). In **Figure 2**, it can be seen that white rot fungi *T. versicolor* almost colonized in cell lumens at 4 weeks due to the obviously present hyphae in the wood cell lumina, and hyphae presented in large clusters after 8 weeks [29]. It can be suggested that white rot fungi *T. versicolor* grew along lumens of wood cell walls during its colonization then the lignin can as well be decomposed accordingly [8, 17]. In brown rot fungi, a large number of cell walls were deconstructed, revealing fungi were able to grow and reproduce, leading to the destruction of wood cell walls. It was also reported that decay fungi colonized and attacked through parenchyma cells via pits and the wood rays were the primary paths for the spread of mycelium [19].
It can be concluded that there also are microstructure changes within wood during the degradation by fungi with the destruction of the cell wall materials, resulting in the enhancement of the accessibility of wood substrates as well as the improvement of the wood permeability.

4. Changes in physical and mechanical properties

Decaying fungi can lead to the changes in the physical and mechanical properties of wood, such as a decrease in intensity, elastic modulus, hardness, etc. Simultaneously, the toughness of wood will also deteriorate. These changes will affect the stability and service life, even the safety of wood products.

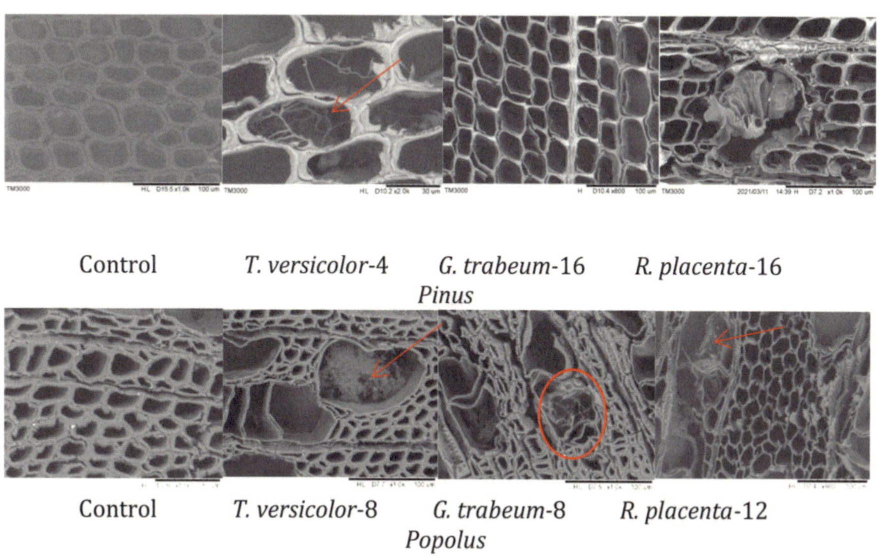

Figure 2.
SEM images of wood samples biodegraded by different fungi.

Wood decay by fungi can cause wood substrate losses, and changes in the chemical compositions, microstructures and physical and mechanical properties of wood. Among those properties, strength is a prior concern when study wood decay due to its critical significance in most structural uses. In summary, the degradation caused by fungi can damage the wood cell walls, leading to the destruction of organizational constituents and structure and the embrittlement of wood, thereby affecting its performance and service life.

3. Wood preservatives

As aforementioned, one of the major drawbacks of wood is its susceptibility to biological deterioration. Wood can be attacked by decay fungi inadvertently in natural conditions and as such its durability is accordingly reduced, accompanied by some decay hazards and others such as losses amounting to billions of dollars each year. It is the primary reason that wood needs to be treated aiming to prolong its service life. Based on the background, the chemical wood preservative and protection technology have been greatly developed worldwide [36].

3.1 Conventional wood preservatives

Generally, conventional wood preservatives can commonly be classified into three types, i.e. oil type wood preservatives, oil-borne type wood preservatives and water-borne type wood preservatives. Coal tar and creosote, which are usually known as traditional oil preservatives, are toxic and effective in resistance against wood decay fungi, insects and other microorganisms. Oil-borne preservatives, which dissolve some toxic water-insoluble organic fungicide compounds in organic solvents [37, 38], are also efficient but limited utilized in certain wood products due to their volatile organic compound problems. During the actual production and application process, some traditional preservatives (oil preservatives or oil-borne preservatives) are poisonous to the environment and human health, due to their containing some toxic chemicals [39]. Therefore, water-borne preservatives have been developed and applied instead of traditional preservatives in many areas.

Water soluble preservatives can protect wood efficiently in most environments, which dissolve some active chemical ingredients in an aqueous solution, which is valid in the inhibition of the harmful microorganisms to wood due to its toxicity to fungi, bacteria, insects, and other biological erosion. Besides, water-borne preservatives can also penetrate into wood cell walls due to their good permeability, and effectively protect the entire wood from the erosion of microorganisms, then provide long-lasting protective effects. Wood water-borne preservatives are usually prepared in liquid form, which can be applied to wood and/or wood surfaces through simple methods such as soaking, spraying, or brushing. Compared with other types of preservatives (oil preservatives or oil-borne preservatives), the application of water-borne preservatives is very convenient. Thus, water-borne preservatives attract increasing attention due to their well-treatment feature, lower toxicity, less environmental impact and pollution after use. Collectively, up to date, water-borne preservatives are the most widely applied wood preservative.

However, there also are some disadvantages to water-borne preservatives, including limited fixation and durability due to their high solubility in water, and low protection compared to some oil preservatives or oil-borne preservatives, especially in extreme environmental conditions, such as high humidity.

3.2 Environmentally benign wood preservatives

Wood is easy to be deteriorated for ubiquitous organisms, such as fungi, bacteria and insects. For this reason, wood products require chemical treatment rather than soil direct contact to prolong their operating lives. Though water-borne preservatives are widely utilized attributing to their inexpensive cost and good permeability, their high solubility also brings negative effects such as low stability and fixation, even risk to the environment due to the toxic leachable chemicals. Chromated copper arsenate (CCA) has been extensively used to effectively protect wood for nearly 100 years. Arsenic and copper compounds are used as toxic elements to the microorganisms and insects in CCA components, while the chromium salt is applied to fixable agents and prevent them from leaching from the CCA-treated lumber into the environment. However, CCA-treated wood products need careful use and cautious disposal for it is a toxic waste and harmful to humans, animals and the environment due to chromium and arsenate in CCA elements are inaccessible to standard toxicity characteristic leaching profile (TCLP) tests [40]. Since 2004 the U.S. Environmental Protection Agency prohibited CCA for residential purposes due to its hazard during manufacture and treatment. As a consequence, it is urgent to develop and research feasible, effective, environment-friendly and cost-competitive wood preservatives to substitute for traditional preservatives, such as CCA.

Under the above background, environmentally benign wood preservatives are researched and developed by many wood science researchers and wood preservative companies during the past few years. Copper salts, which are poisonous to microorganisms and insects, have been used most frequently in wood preservatives and could react with and/or bind to lignin, tannin, or protein consequently fixed in the wood. Boron salts are the oldest preservatives and are still used as effective fungicides and insecticides nowadays on account of their low toxicity [41, 42]. Due to the preservative active ingredients and low toxicity, copper and boron salts attracted more and more attention [43–45]. However, the leachates (copper and/or boron elements) from the treated products during the long-time application are inevitable for their water solubility [14]. Thus, copper and/or boron-based wood preservatives need to be developed to stabilize and fix the active ingredients (copper and/or boron elements) onto wood structures.

Recently, proteins such as soy isolates, okara protein, and feather protein, have been used to interact with the preservative active ingredients by coagulation, auto-condensation, and/or other chemical reactions to increase the durability of the preservative in treated wood [36, 46]. When the copper-boron-protein preservatives impregnate the wood, copper and boron can interact with wood components and be fixed in the wood matrix by gelling of protein via heating or other methods. Mazela et al. (2003) and Thevenon et al. (1998) used proteins and tannin compounds to fix the boric acid during two impregnation stages [47–49]. Sye et al. (2008) prepared a wood preservative by formulating copper and/or borax with organic waste okara to substitute the high-price copper azoles (CuAz) and alkaline copper quaternary (ACQ) [50]. Yang (2006) studied the feasibility of using soy protein instead of toxic chromium and arsenic to formulate wood preservatives with copper and boron [51]. The aforementioned studies proved that the protein-based wood preservatives could penetrate the wood block and protect the wood products against fungal attack as effectively as traditional preservatives, such as CuAz. In addition, they are environmentally friendly and have been considered as an interest alternative to CCA.

Based on this theory, the authors developed a kind of environmental friendly wood preservative based on chicken feather protein, which was used as the source of protein for its environmental benign character and low cost. The preservative formulations were composed of hydrolyzed chicken feather protein, copper sulfate ($CuSO_4 \cdot 5H_2O$) and sodium borate ($Na_2B_4O_7 \cdot 10H_2O$). Chicken feather powder was hydrolyzed at 140°C for 4 h, then the protein hydrolysate was obtained. The condensed hydrolyzate was added into the suspension of copper sulfate and sodium borate. Then the wood preservative solution based on feather protein was achieved with the dissolving agent ammonium hydroxide (NH_4OH) [36].

The results showed that chicken feather proteins can be successfully used to prepare the protein-based wood preservative, which can penetrate wood structures and are stable against water leaching. The interactions between chicken feather protein-based wood preservatives and wood components were also confirmed. Therefore, chicken feather protein could be used as a source of protein and an efficient chelating agent to prepare low-cost, effective and environmentally benign wood preservatives, and the chicken feather protein-based preservative can effectively protect the wood against decay fungi and prolong the service life of the treated wood blocks, which provides a new source of protein using natural components as potential wood preservatives. For exploring the ground-contact protection of the chicken feather protein-based preservative, field trials with much longer processing time need to be conducted in the future in order to evaluate the long-term effectiveness of this kind of preservative.

Natural materials attract more and more interest as a source of preservatives due to their simple way to obtain, low cost and environmentally friendly characteristics. There are some new preservatives prepared from natural materials, due to their competitive cost, low toxicity and low environmental impact, such as plant-derived wood preservatives. For instance, Tiina Belt investigated the extractives of heartwood of Scots pine, containing extractives, such as pinosylvins, and suggested that pine heartwood extractives have the potential to inhibit the white rot fungi [52]. The ethanol extractives of teak heartwood residues also showed promising antifungal abilities as wood preservatives [6]. Tchinda reported that plant essential oil showed positive antifungal activities, using natural plant extracts to protect wood [38, 53]. Senmiao Fang mixed chitosan and cinnamaldehyde as a kind of natural wood preservative and proved that the new kind of preservative can effectively protect the test sample, which can be easily used and overcomes the volatilization problem. Salicylic acid, also a natural organic substance extracted from plants, possessing antibacterial functions, can also be utilized and formulated as a kind of wood preservative. Li Yan formulated a salicylic acid/silica microcapsule and studied its decay resistance as well as the stability of modified poplar wood. The decay resistance of treated poplar was greatly improved compared to untreated poplar [54].

4. Conclusions

Wood is susceptible to being infected, decayed and deteriorating in the natural environment due to the ubiquitous microorganisms. Many microorganisms, such as wood decay fungi and bacteria, could attack and damage wood products leading to enormous commercial waste and property losses. Wood decay, an inevitable natural phenomenon caused by the activity of some microorganisms, brought about various hazards, such as wood structural deformation, and the losses of original strength and stability, which could lead to deconstruction and collapse of wooden products, causing economic impact and property damage even harm and danger to citizens. The decay mechanisms of wood are different depending on the different decay organisms (fungi, bacteria or insects) and various wood species including diverse decay stages. It is very important to understand the decay mechanisms of wood. Through the elucidation of the decay mechanisms of wood, scientific protection technology or measures could be performed in order to prolong the service life of wood products, which is crucial both for sustainable forest resource management and the research and development of wood preservatives. To enhance the resistance of wood against the decay fungi or bacteria, chemical treatments and preservatives have frequently been applied in the wood industry. Environmental concerns have prompted the development of wood preservatives based on natural materials, which are with high efficacy, low cost, low health risks and low environmental impacts. Some protein-based wood preservatives or wood extractives have received great attention due to their low cost, toxicity and environmental impact. As scientific research and technologies advance, the decay mechanism of wood and its relationship with wood preservatives have been further developed, providing an improved understanding of the wood degradation process by various microorganisms (fungi, bacteria) and promoting scientific wood protection and maintenance.

Acknowledgements

The authors would like to express their sincere gratitude to Prof. Mizi Fan and Dr. Yonghui Zhou from the Department of Civil and Environmental Engineering at Brunel University London for their valuable suggestions in reviewing the manuscript.

The authors also are grateful for the financial support by the Regional Project of National Natural Science Foundation of China (32260362, 31860186, 31360157), the Joint project of Yunnan Agricultural Basic Research (202101BD070001-058), and the 111 Project (D21027).

Conflict of interest

The authors declare no conflict of interest.

Author details

Yan Xia[1,2*] and Lu Jia[1]

1 Yunnan Provincial Key Laboratory of Wood Adhesives and Glued Products, Southwest Forestry University, Kunming, China

2 Department of Civil and Environmental Engineering, Brunel University London, Uxbridge, UK

*Address all correspondence to: xiayanswfu@gmail.com

References

[1] Broda M, Popescu C. Natural decay of archaeological oak wood versus artificial degradation processes — An FT-IR spectroscopy and X-ray diffraction study. Spectrochimica Acta. Part A, Molecular and Biomolecular Spectroscopy. 2019;**209**:280-287. DOI: 10.1016/j.saa.2018.10.057

[2] Blanchette RA. A review of microbial deterioration found in archaeological wood from different environments. International Biodeterioration & Biodegradation. 2000;**46**:189. DOI: 10.1016/s0964-8305(00)00077-9

[3] Yu H, Guo G, Zhang X, Yan K, Xu C. The effect of biological pretreatment with the selective white-rot fungus Echinodontium taxodii on enzymatic hydrolysis of softwoods and hardwoods. Bioresource Technology. 2009;**100**:5170-5175. DOI: 10.1016/j.biortech.2009.05.049

[4] Li T, Cui L, Song X, Cui X, Wei Y, Tang L, et al. Wood decay fungi: An analysis of worldwide research. Journal of Soils and Sediments. 2022;**22**:1688-1702. DOI: 10.1007/s11368-022-03225-9

[5] Fernandes L, Loguercio-Leite C, Esposito E, Menezes Reis M. In vitro wood decay of Eucalyptus grandis by the basidiomycete fungus Phellinus flavomarginatus. International Biodeterioration & Biodegradation. 2005;**55**:187-193. DOI: 10.1016/j.ibiod.2004.12.001

[6] Brocco VF, Paes JB, Costa LGD, Brazolin S, Arantes MDC. Potential of teak heartwood extracts as a natural wood preservative. Journal of Cleaner Production. 2017;**142**:2093-2099. DOI: 10.1016/j.jclepro.2016.11.074

[7] Riley R, Salamov AA, Brown DW, Nagy LG, Floudas D, Held BW, et al. Extensive sampling of basidiomycete genomes demonstrates inadequacy of the white-rot/brown-rot paradigm for wood decay fungi. Proceedings of the National Academy of Sciences - PNAS. 2014;**111**:9923-9928. DOI: 10.1073/pnas.1400592111

[8] Skyba O, Douglas CJ, Mansfield SD. Syringyl-rich lignin renders poplars more resistant to degradation by wood decay fungi. Applied and Environmental Microbiology. 2013;**79**:2560-2571. DOI: 10.1128/AEM.03182-12

[9] Zabel RA, Morrell JJ. The characteristics and classification of fungi and bacteria. In: Wood Microbiology the Characteristics and Classification of Fungi and Bacteria. United States: Elsevier Science & Technology; 2020. p. 55

[10] Siitonen J. Forest management, coarse Woody debris and Saproxylic organisms: Fennoscandian boreal forests as an example. Ecological Bulletins. 2001;**49**:11-41

[11] Zabel RA, Morrell JJ. Factors affecting the growth and survival of fungi in wood (fungal ecology). In: Wood Microbiology Factors Affecting the Growth and Survival of Fungi in Wood (Fungal Ecology). United States: Elsevier Science & Technology; 2020. p. 99

[12] Rytioja J, Hildén K, Yuzon J, Hatakka A, de Vries RP, Mäkelä MR, et al. Plant-polysaccharide-degrading enzymes from basidiomycetes. Microbiology and Molecular Biology Reviews. 2014;**78**:614-649. DOI: 10.1128/MMBR.00035-14

[13] Arantes V, Goodell B. Current understanding of Brown-rot fungal

biodegradation mechanisms: A review. ACS Symposium Series. 2014;**3**:3-21. DOI: 10.1021/bk-2014-1158.ch001

[14] Liu M, Zhong H, Ma E, Liu R. Resistance to fungal decay of paraffin wax emulsion/copper azole compound system treated wood. International Biodeterioration & Biodegradation. 2018;**129**:61-66. DOI: 10.1016/j.ibiod.2018.01.005

[15] Arantes V, Milagres AMF, Filley TR, Goodell B. Lignocellulosic polysaccharides and lignin degradation by wood decay fungi: The relevance of nonenzymatic Fenton-based reactions. Journal of Industrial Microbiology & Biotechnology. 2011;**38**:541-555. DOI: 10.1007/s10295-010-0798-2

[16] Hastrup ACS, Howell C, Larsen FH, Sathitsuksanoh N, Goodell B, Jellison J. Differences in crystalline cellulose modification due to degradation by brown and white rot fungi. Fungal Biology. 2012;**116**:1052-1063. DOI: 10.1016/j.funbio.2012.07.009

[17] Bari E, Daryaei MG, Karim M, Bahmani M, Schmidt O, Woodward S, et al. Decay of Carpinus betulus wood by Trametes versicolor - an anatomical and chemical study. International Biodeterioration & Biodegradation. 2019;**137**:68-77. DOI: 10.1016/j.ibiod.2018.11.011

[18] Qi J, Jia L, Liang Y, Luo B, Zhao R, Zhang C, et al. Fungi's selectivity in the biodegradation of Dendrocalamus sinicus decayed by white and brown rot fungi. Industrial Crops and Products. 2022;**188**:115726. DOI: 10.1016/j.indcrop.2022.115726

[19] Qi J, Zhang X, Zhou Y, Zhang C, Wen J, Deng S, et al. Selectively enzymatic conversion of wood constituents with white and brown rot fungi. Industrial Crops and Products. 2023;**199**:116703. DOI: 10.1016/j.indcrop.2023.116703

[20] Kahl T, Arnstadt T, Baber K, Bässler C, Bauhus J, Borken W, et al. Wood decay rates of 13 temperate tree species in relation to wood properties, enzyme activities and organismic diversities. Forest Ecology and Management. 2017;**391**:86-95. DOI: 10.1016/j.foreco.2017.02.012

[21] Beck G, Thybring EE, Thygesen LG. Brown-rot fungal degradation and de-acetylation of acetylated wood. International Biodeterioration & Biodegradation. 2018;**135**:62-70. DOI: 10.1016/j.ibiod.2018.09.009

[22] Ray MJ, Leak DJ, Spanu PD, Murphy RJ. Brown rot fungal early stage decay mechanism as a biological pretreatment for softwood biomass in biofuel production. Biomass & Bioenergy. 2010;**34**:1257-1262. DOI: 10.1016/j.biombioe.2010.03.015

[23] Yelle DJ, Ralph J, Lu F, Hammel KE. Evidence for cleavage of lignin by a brown rot basidiomycete. Environmental Microbiology. 2008;**10**:1844-1849. DOI: 10.1111/j.1462-2920.2008.01605.x

[24] Gabriel J, Švec K. Occurrence of indoor wood decay basidiomycetes in Europe. Fungal Biology Reviews. 2017;**31**:212-217. DOI: 10.1016/j.fbr.2017.05.002

[25] Eriksson K-EL, Blanchette RA, Ander P. Morphological aspects of wood degradation by fungi and bacteria. In: Microbial and Enzymatic Degradation of Wood and Wood Components. Heidelberg: Springer Berlin; 2012. pp. 1-87

[26] Zabel RA, Morrell JJ. Chemical changes in wood caused by decay fungi. In: Wood Microbiology Chemical

Changes in Wood Caused by Decay Fungi. United States: Elsevier Science & Technology; 2020. p. 215

[27] Zabel RA, Morrell JJ. Changes in the strength and physical properties of wood caused by decay fungi. In: Wood Microbiology Changes in the Strength and Physical Properties of Wood Caused by Decay Fungi. United States: Elsevier Science & Technology; 2020. p. 271

[28] Wilcox WW. Changes in Wood Microstructure through Progressive Stages of Decay. Madison, Wisconsin: U. S. Department of Agriculture/Forest Service/Forest Products Laboratory; 1968. pp. 1-48

[29] Qi J, Li F, Jia L, Zhang X, Deng S, Luo B, et al. Fungal selectivity and biodegradation effects by white and Brown rot fungi for wood biomass Pretreatment. Polymers. 2023;**15**. DOI: 10.3390/polym15081957

[30] Brischke C, Hanske M. Durability of untreated and thermally modified reed (Phragmites australis) against brown, white and soft rot causing fungi. Industrial Crops and Products. 2016;**91**:49-55. DOI: 10.1016/j.indcrop.2016.06.031

[31] Zhenfu L, Biao P, Buyun L, Xuefeng Z, Shiyu C. Density and shrinkage of poplar wood grown on the riverside of Yangtze River. Forestry Science and Technology Development. 2012;**2**:60-62

[32] Shutong F, Jihang H, Junliang L, Shuang DH. Comparison of properties of rubber wood from two different producing areas and preliminary modification. Journal of Beihua University (Natural Science Edition). 2019;**2**:256-259

[33] Lihong Q, Xiaoling L, Liufeng L, Yunlin F, Mei L. Green wood properties of pinus yunnanensis var. tenuifolia. Journal of Northwest Forestry University. 2015;**3**:7

[34] Nian T, Xueshun W, Anmin H, Chen W. Wood density prediction of Cunninghamia lanceolata based on Gray wolf algorithm SVM and NIR. Forestry Science. 2018;**12**:5

[35] Sunardi, Tanabe J, Ishiguri F, Ohshima J, Iizuka K, Yokota S. Changes in lignocellulolytic enzyme activity during the degradation of Picea jezoensis wood by the white-rot fungus Porodaedalea pini. International Biodeterioration & Biodegradation. 2016;**110**:108-112. DOI: 10.1016/j.ibiod.2016.02.022

[36] Xia Y, Ma C, Wang H, Sun S, Wen J, Sun R. Multiple analysis and characterization of novel and environmentally friendly feather protein-based wood preservatives. Polymers. 2020;**12**:1-14. DOI: 10.3390/POLYM12010237

[37] Nguyen TTH, Li J, Li S. Effects of water-borne rosin on the fixation and decay resistance of copper-based preservative treated wood. BioResources. 2012;7:3573-3584

[38] Fang S, Feng X, Lei Y, Chen Z, Yan L. Improvement of wood decay resistance with cinnamaldehyde chitosan emulsion. Industrial Crops and Products. 2021;**160**:113118. DOI: 10.1016/j.indcrop.2020.113118

[39] Medeiros FCMD, Gouveia FN, Bizzo HR, Vieira RF, Del Menezzi CHS. Fungicidal activity of essential oils from Brazilian Cerrado species against wood decay fungi. International Biodeterioration & Biodegradation. 2016;**114**:87-93. DOI: 10.1016/j.ibiod.2016.06.003

[40] Thevenon M-F, Pizzi A, Haluk J-P. Non-toxic albumin and soja protein borates as ground-contact wood preservatives. European Journal of Wood and Wood Products. 1997;**55**:293-296. DOI: 10.1007/s001070050231

[41] Baysal E, Ozaki SK, Yalinkilic MK. Dimensional stabilization of wood treated with furfuryl alcohol catalysed by borates. Wood Science and Technology. 2004;**38**:405-415. DOI: 10.1007/s00226-004-0248-2

[42] Garcia R, Gano L, Maria L, Paulo A, Santos I, Spies H. Synthesis and biological evaluation of tricarbonyl Re(I) and Tc(I) complexes anchored by poly(azolyl)borates: Application on the design of radiopharmaceuticals for the targeting of 5-HT1A receptors. Journal of Biological Inorganic Chemistry. 2006;**11**:769-782. DOI: 10.1007/s00775-006-0124-7

[43] Kartal SN, Hwang W, Shinoda I, Y. Laboratory evaluation of boron-containing quaternary ammonia compound, didecyl dimethyl ammonium tetrafluoroborate (DBF) for control of decay and termite attack and fungal staining of wood. European Journal of Wood and Wood Products. 2006;**64**:62-67. DOI: 10.1007/s00107-005-0050-3

[44] Nguyen TTH, Li S, Li J, Liang T. Micro-distribution and fixation of a rosin-based micronized-copper preservative in poplar wood. International Biodeterioration & Biodegradation. 2013;**83**:63-70. DOI: 10.1016/j.ibiod.2013.02.017

[45] Humar M, Thaler N. Performance of copper treated utility poles and posts used in service for several years. International Biodeterioration & Biodegradation. 2017;**116**:219-226. DOI: 10.1016/j.ibiod.2016.11.004

[46] Polus-Ratajczak I, Mazela B. The use of blood protein in wood preservatives. European Journal of Wood and Wood Products. 2004;**62**:181-183. DOI: 10.1007/s00107-004-0477-y

[47] Mazela B, Polus-Ratajczak I. Use of animal proteins to limit leaching of active copper ions preservatives from treated wood. Holzforschung. 2003;**57**:593-596. DOI: 10.1515/HF.2003.089

[48] Thevenon MF, Pizzi A, Haluk J, Zaremski A. Normalised biological tests of protein borates wood preservatives. European Journal of Wood and Wood Products. 1998;**56**:162. DOI: 10.1007/s001070050290

[49] Lyon F, Thevenon M, Hwang W, Imamura Y, Gril J, Pizzi A. Effect of an oil heat treatment on the leachability and biological resistance of boric acid impregnated wood. Annals of Forest Science. 2007;**64**:673-678. DOI: 10.1051/forest:2007046

[50] Ahn SH, Oh SC, Choi I-G, Kim H-Y, Yang I. Efficacy of wood preservatives formulated from okara with copper and/or boron salts. Journal of Wood Science. 2008;**54**:495- 501. DOI: 10.1007/s10086-008-0982-4

[51] Yang I, Kuo M, Myers DJ. Soy protein combined with copper and boron compounds for providing effective wood preservation. Journal of the American Oil Chemists' Society. 2006;**83**:239-245. DOI: 10.1007/s11746-006-1199-6

[52] Belt T, Mollerup F, Hänninen T, Rautkari L. Inhibitory effects of scots pine heartwood extractives on enzymatic holocellulose hydrolysis by wood decaying fungi. International Biodeterioration & Biodegradation. 2018;**132**:150-156. DOI: 10.1016/j.ibiod.2018.03.004

[53] Saha Tchinda J, Ndikontar MK, Fouda Belinga AD, Mounguengui S, Njankouo JM, Durmaçay S, et al. Inhibition of fungi with wood extractives and natural durability of five Cameroonian wood species. Industrial Crops and Products. 2018;**123**:183-191. DOI: 10.1016/j.indcrop.2018.06.078

[54] Yan L, Zeng F, Chen Z, Chen S, Lei Y. Improvement of wood decay resistance by salicylic acid / silica microcapsule: Effects on the salicylic leaching, microscopic structure and decay resistance. International Biodeterioration & Biodegradation. 2021;**156**:105134. DOI: 10.1016/j.ibiod.2020.105134

Chapter 6

Biopolymers as Coating Additives for Engineered Wood Products

Mihaela Tanase-Opedal

Abstract

Engineered wood products are used as a construction material due to enhance performance, faster and higher construction of buildings, durability, and less impact on the environment. However, its flammability and resistance to mold, insects and water limits its use in construction, and especially in exterior use. Thus, the necessity of developing wood coating formulations that reduce the impact of the environment and increases the durability of engineered wood products. Biopolymers have attracted considerable interest as alternatives in coating applications for engineered wood products due to their availability, environmentally friendly and compatibility with the main wood components. The focus of this book chapter is to give an overview of the treatment methods and bio-based coating of the engineered wood, with special emphasis on lignin-based coating. Lignin/lignin nanoparticles, due to the presence of functional groups, is a promising polymer for coating formulations and applications. Lignin can produce a significant quantity of char when heated at high temperatures. This is important combustion characteristic when lignin is to be used as coating additive for wood. As such, lignin-based fire retardant and antibacterial action of lignin are important properties when lignin-based coating formulations are developed, and they are discussed in this chapter.

Keywords: biopolymers, engineered wood products, lignin-based coating, lignin-fire retardant, antibacterial activity of lignin

1. Introduction

Engineered wood products are building materials that are made by laminating layers of wood together or by binding wood fibers together into a composite material, typically with an adhesive usually involving heat and/or pressure [1]. Engineered wood products offer consistency of structural performance and dimensional stability, making it possible to integrate them successfully with other construction materials on large and complex projects [2]. Engineered wood products are making it conceivable to build taller and bigger wood structures, which is highly asked by the building market [3]. There are many different types of engineered wood products, which can be categorized according to the type of feedstock used in their manufacture. The engineered wood association offer guidance on the properties and applications of engineered wood products being classified in three primary categories, as shown in **Figure 1**. Advantages of using engineered wood products compared to alternative

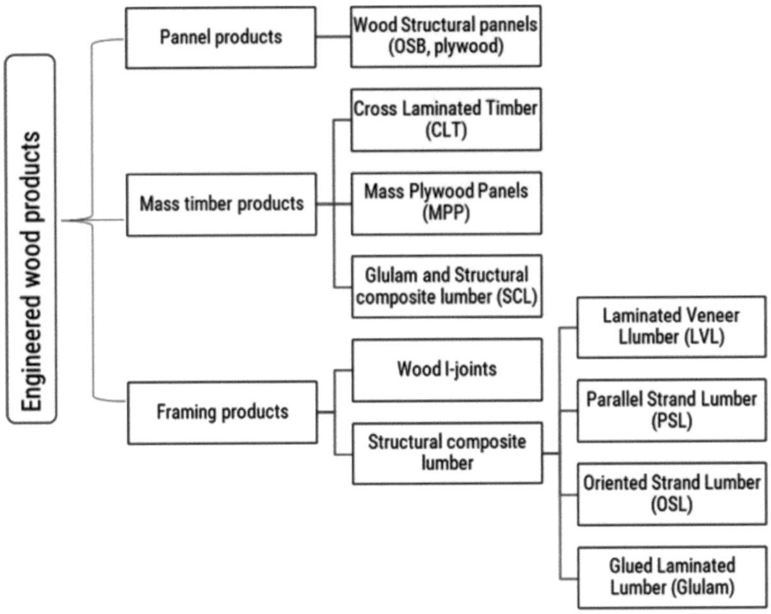

Figure 1.
The classification of engineered wood products, according to APA.

building materials are: (i) natural, renewable, sustainable and reduced carbon footprint; (ii) lighter-weight and greater flexibility in design and construction; (iii) faster and quitter construction; (iv) cheaper construction, warmth and esthetically pleasing with health benefits. Engineered wood products are considered to be renewable construction materials due to their composition [2].

A major factor supporting growth in the use of engineered wood products in construction is the increased environmental and sustainability concern [3, 4], which is influencing construction techniques and the choice of building materials. As such, wood is a renewable and sustainable building material used for modern engineering solutions and functional and decorative applications [5]. Wood has better insulating properties and a positive carbon balance compared to other building materials. Wood has much smaller carbon footprint than other construction materials, and increased use may reduce CO2 emissions by 14% [6]. Moreover, the greenhouse gas emissions have been shown to be as the same level as of concrete and lower than steel [7]. If properly maintained, wood can storage carbon for long lifespan. The engineered wood products used in exterior applications are maintained in time due to the use of preservatives such as creosote, halogenated carbamates, benzothiazoles, pentachlorophenol, (alkyl) imidazoles, bis(tributyltin) oxide, or salt-based impregnates [8, 9]. In addition, fire retardancy is an important behavior if the engineered wood products that are used in construction applications. Currently, the halogenated products are used as fire retardant additives for wood applications [10–13].

Both the currently used preservatives and halogenated fire-retardant additives used today are considered toxic, both for humans and environment. Preservative leaching problem when wood is encountering water and recyclability of wood materials when preservatives are used have been discussed by several authors [14–16]. As such, these concerns contribute to the necessity of developing non-toxic bio-based alternatives. Issues of sustainability and carbon sequestration opens the possibility for new green technologies which can improve durability, stability, and performance of wood, especially in exterior applications [17]. The wood preservative industry is interested in finding low-cost, environmentally friendly methods for the treatment of the wood [18]. Engineered wood products (EWP) producers recognize the potential to greatly expand their market-share if more optimal and affordable wood protection options can be found [1].

Extending the service life of wood and wood-derived products by using environmentally friendly biopolymers represents an attractive approach for wood protection from the perspectives of human health and environmental protection [19, 20]. The modern coating market is dominated by acrylic, polyurethane, and polyester polymer resins produced from unsustainable fossil resources. These coating additives are still used on the market due to their properties and low price. In the recent years, sustainable solutions such as vegetable-oil based coatings such as tall linseed, coconut, soybean, and castor-oil have been introduced on the market [21, 22] These oils are often used in different coating combinations to improve their properties. Usually, most of the additives used in coating applications requires chemical modification [21–24].

An environmentally friendly solution for wood preservation could be the use of biopolymers [25]. Biopolymers, due to the compatibility with the main wood components are considered as interesting alternatives to be used in coating applications [20]. Biopolymers not only that can enhance the performance of adhesives derived from petroleum in different ways [26], but also they can be used to develop environmental friendly and sustainable bio-based alternatives. Biopolymers are categorized as a function of their monomer unit in polysaccharides (cellulose, hemicellulose, glucans, starch), proteins (gelatin, casein), derived polypeptides (collagen, peptides) and polyphenols (tannins, lignin) [25, 27]. Biopolymer-based coatings can be directly deposited onto the substrate surface or by chemical reactions between the biopolymer and the substrate. The chosen coating technique is decided by a specific application, thus depending on several factors, described by Song et al. [28]. Biopolymers, due to the superior compatibility with the main components of wood, have a positive effect on the penetrability of biopolymer into the wooden mass and can enhance the biopolymer biocidal activity. Many other benefits, such as wood recyclability have been pointed out in the work of Patachia and Croitoru [29]. As such, this book chapter is important as it gives an overview on the use of biopolymers in coatings formulations and how these formulations can protect the wood against fire, insects, mold and water. Using natural biopolymers to replace the conventional preservatives and fire-retardant additives in wood protection is highly recommended, as the engineered wood product will be completely renewable and recyclable.

2. Treatment methods of engineered wood products

The drivers of using biopolymers-based additives in wood coating applications consist of sustainable concerns, such as increasing the durability of engineered wood products without the use of toxic compounds. In general, as described in the introduction part, wood can be protected against fungi, molds or insects by impregnation using different natural substances [30, 31]. Impregnation has been mostly applied on the solid wood, where the wood structure was chemically modified so that water penetration in the wood structure was limited. Chemical structure and composition of the engineered wood products allows efficient chemical modifications both at surface and inside the wood structure [32, 33]. As such, there is possible to tailor and synthesis recyclable and renewable engineered wood products with specific properties [18]. The uses of engineered wood products for exterior applications, such as in house-holds buildings requires both a surface and chemical treatment. Exterior wood coatings represent the second largest segment accounting 25% of the global architectural wood coatings market [7]. Different coatings formulations are designed to protect the wood from weathering degradation and preservation in outdoor conditions [34]. The coating agents act at the wood surface as barriers against environmental factors action, such as attack of insects, moisture and fire) and to maintain the aesthetical appearance of the wood. Plant oil-type wood preservatives, such as wood and plants extractives, vegetable oils, natural waxes, different biopolymers and biological control agents are the most applied one [31].

Both, high and low molecular mass biopolymers protect the wood against moisture, oxygen and biological attack. However, the mechanism between low and high molecular mass biopolymers is different. Biopolymers with high molecular mass can be used for surface impregnation, forming viscous biopolymer solutions which can minimize the leaching of biocidal compounds from the treated wood [35]. As such, biopolymer coating formulations can protect the environment and prolonging the lifecycle of wood [35]. Natural biopolymers, with low molecular mass generate solutions (aqueous or organic solvent based) with low viscosity, are proposed to be used as impregnation agents by diffusion into the wood. As such, by creating a film inside wood lumen and closing the pores, allows protection of the treated engineered wood products against water and biological attack. These low-molecular biopolymer solutions could be introduced into wood either by immersing (superficial impregnation) or by high-pressure impregnation [36, 37]. As such, enzymatic polymerization of essential oils with lignin in wood and treatments with nanoparticles [18], represents a promising solution to the engineered wood treatment as illustrated in **Figure 2**.

Currently, biopolymers are used in wood impregnation as aqueous dispersion or emulsions. However, new techniques for using biopolymers in wood modifications are developed in the last years such as biopolymer hydrogels, nanoparticles or biopolymer insertion by using an organic solvent as carrier. The hydrogels or the nanoparticles

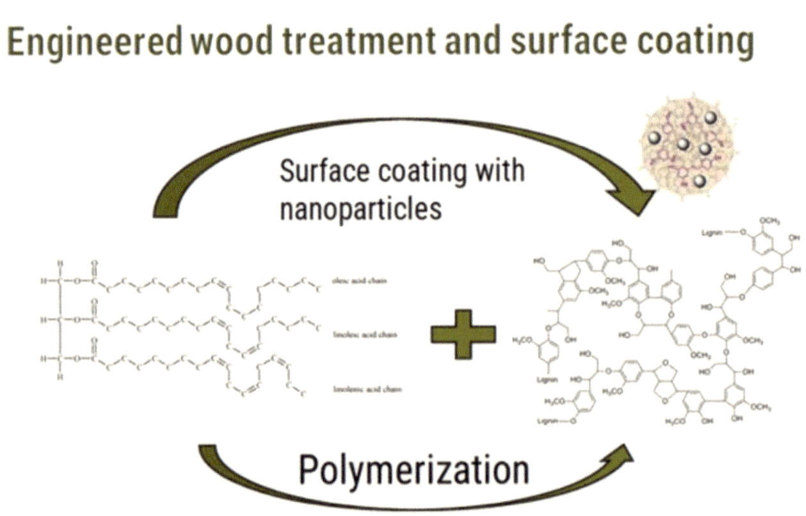

Figure 2.
Schematic figure of engineered wood treatment and surface coating.

can be loaded with biocides and within controlled conditions of temperature and moisture favors the swelling and the diffusion of hydrogels into the holes in the wood structure. As such, the biocide is fixed in the wood structure avoiding the leaching problem. When an organics solvent is used as carrier for the biopolymer, the same swollen mechanism was observed [38–40].

3. Bio-based coatings

In the last years, have bio-based adhesives gained considerable interest in the bonding of engineered wood products, as environmentally friendly approach compared to the conventional based adhesives. Natural biopolymers such as, cellulose, protein, lignin, and tannin and their modifications with different dispersing agents and cross-linkers have been successfully applied as adhesives for bonding of the engineered wood products [41]. Because of its hydrophobicity, lignin can be used as raw materials for coating [42]. Lignin, due to the presence of phenol groups in its structure, can successfully replace phenol in lignin-based adhesives formulations. Siahkamari et al. [43] developed a bio-based phenolic adhesive by entirely substituting both fossil-based phenol and formaldehyde with lignin and glyoxal.

3.1 Lignin-based coatings

Lignin, as a natural biopolymer from wood is produced as a by-product in many biorefinery processes. Currently, only about 1 million ton is used for value-added purposes, which mainly comprise in dispersants, adhesives, and fillers [44, 45]. Lignin has a complex chemical structure which includes hydroxyl, carboxyl and phenolic groups. The presence of these groups depends on the lignin isolation process. Chemical/enzymatic modification of lignin is often a necessity to introduce new functional groups that will increase compatibility between the components in the final material, illustrated in **Figure 3**. Interest in substituting fossil-based polymers with biopolymers in coating industry represents a great market opportunity in channeling recent developments into the production of green coating additives for engineered wood products. Lignin conversion to high quality products is critical to a biorefinery's profitability and sustainability. Organosolv lignin was esterified using dodecanoyl chloride to synthesize a hydrophobic coating for wood [46]. Literature studies show that lignin-based coatings have improved water repellent properties compared to conventional coatings formulations [46]. Henn et al. [47] demonstrated the preparation of fully particulate coatings without the use of binding matrix using lignin instead of metal oxides. Furthermore, colloidal lignin particles were exploited to prepare water-based, solvent free, and multiresistant surface coatings. Due to their hydroxyl groups, the colloidal lignin particles acted as hardener and required no binder to adhere to the substrate. As such, organosolv lignin has been successfully employed to prepare lignin-based epoxy resins [48].

Micro- and nanostructured coatings, such as colloidal lignin particles or lignin nanoparticles have gained attention because they disperse easy and due to their often excellent anticorrosion, antibacterial, anti-icing, and UV-shielding properties [49, 50]. It has been shown that high surface roughness of nanostructured coatings is one important factor contributing to their exceptional hydrophobicity [51, 52]. Hydrophobicity and abrasion of nanostructured coatings can be improved by binding or encapsulating the particles to a polymer/biopolymer matrix obtaining in this way a covalently particle-polymer matrix. As such, particle-polymer matrix can be applied

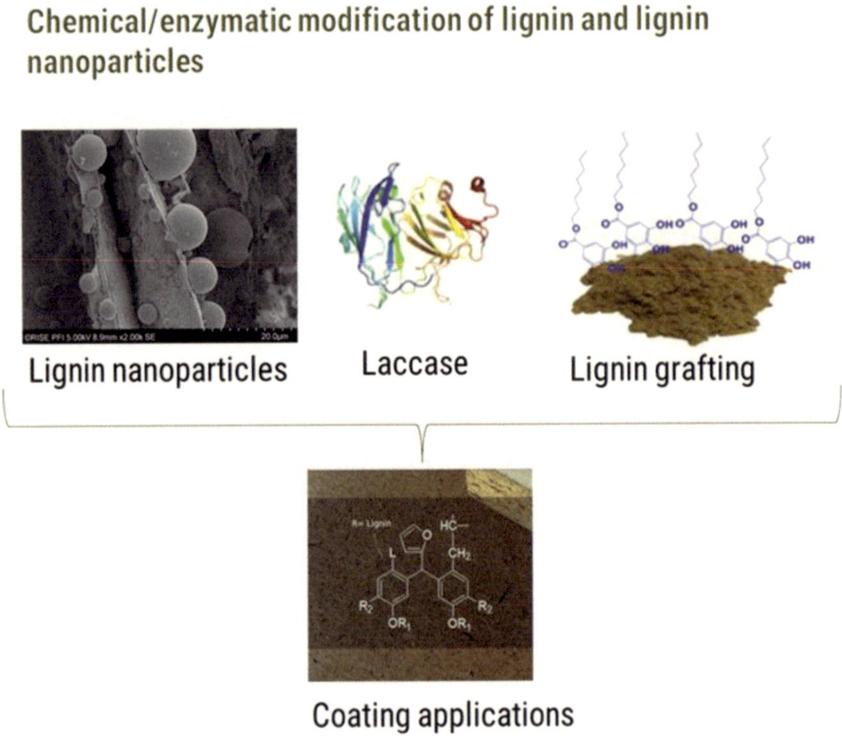

Figure 3.
Illustration of lignin nanoparticles, chemical/enzymatic modification of lignin.

for special applications being shown in the literature to give a very good water and abrasion protection, but at higher price [50, 53–55].

3.1.1 Lignin-based fire retardant

In recent research literature, it has been found that intumescent flame retardants, such as ammonium polyphosphate (APP), as a non-reactive, inorganic material can be added to polymers as a substitute for halogen flame retardant, being compatible with many polymers and biopolymers [56]. APP is a reaction results of ammonia and phosphoric acid. Therefore, as an additive used for intumescent coating in flame-retardant applications, APP has both function of acid and gas source. When a product containing APP meets fire, APP acts as a flame retardant by a chemical effect in the condensed phase called intumescence. As a result, a carbon foam is formed at the surface of the material which acts as an insulating layer, preventing further decomposition of the material. It has been shown that APP has high content of phosphorus and nitrogen, environmentally friendly, good thermal stability, low smoke, and nontoxicity [11]. These characteristics makes the intumescent flame superior to conventional flame retardants. In recent years, lignin and chitosan has been used as a carbon source in different flame-retardant formulations [12, 13, 57].

Lignin based flame retardants can be prepared by directly physical blending or by chemical modification [58, 59]. A disadvantage of the physical method is

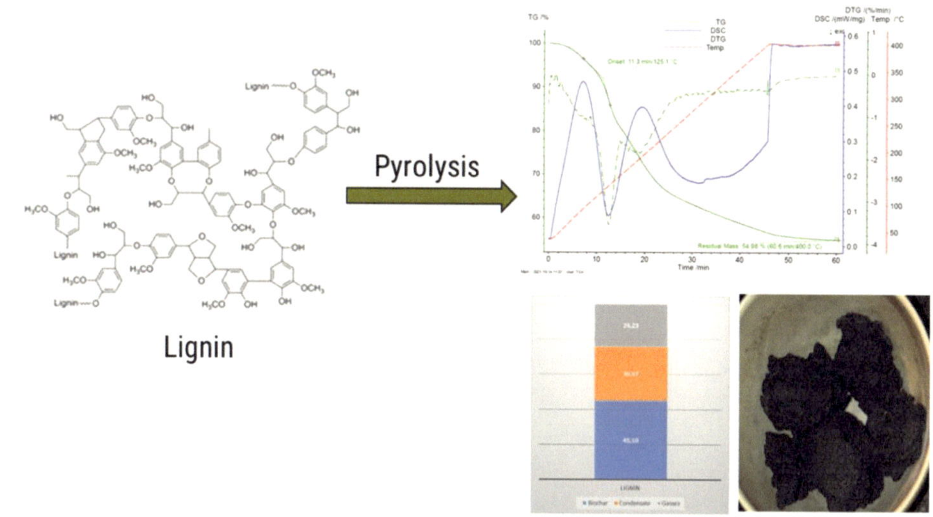

Figure 4.
Thermal decomposition of lignin.

uneven multicomponent mixing, which has a negative impact on flame retardancy. During the chemical modification, the hydroxyl groups present in lignin structure will react with desired functional groups, such as ammonia, phosphoric acid given a lignin with modified structure which is suitable to be used in intumescent flame-retardant formulations [60–63]. Zhang et al. [64] showed that lignin modified with urea and combined it with ammonium polyphosphate (APP) was successfully used as a novel intumescent flame retardant (IFR) system to improve the flame retardancy of polylactic acid (PLA). Moreover, Liu et al. [63] showed that novel lignin-based flame retardant was done by chemically grafting nitrogen, phosphorus and copper elements into lignin structure to improve the flame retardance of wood- plastic composites. Lignin nanoparticles can also be used in different intumescent flame-retardant formulations. Collet et al. used for the first-time lignin nanoparticles modified with phosphor in intumescent flame-retardant formulations [65].

Char yield during combustion of a polymer is an important characteristic when the polymer is to be used as a flame retardant or as additive for intumescent coating. We have observed that during thermal decomposition lignin produces high char yield up to 45%, as seen in **Figure 4**. We believe that the high char layer has a positive effect on smoke suppression and therefore er lignin a promising additive in intumescent flame-retardant formulations.

Our hypothesis on formation of a larger and denser charring layer helps in improving smoke suppression is in accordance with literature results of Dai et al. [59]. The mechanism involved here is similar when APP is used, where the hydroxyl groups present in lignin structure reacts with phosphoric acid and ammonia, as illustrated in **Figure 5.**

As such, both phosphor and nitrogen are introduced in lignin structure, having a function of an acid and gas source in intumescent flame-retardant formulations. We strongly believe that the synergic effect of both nitrogen and phosphor incorporated in lignin structure can improve the fire-resistance properties.

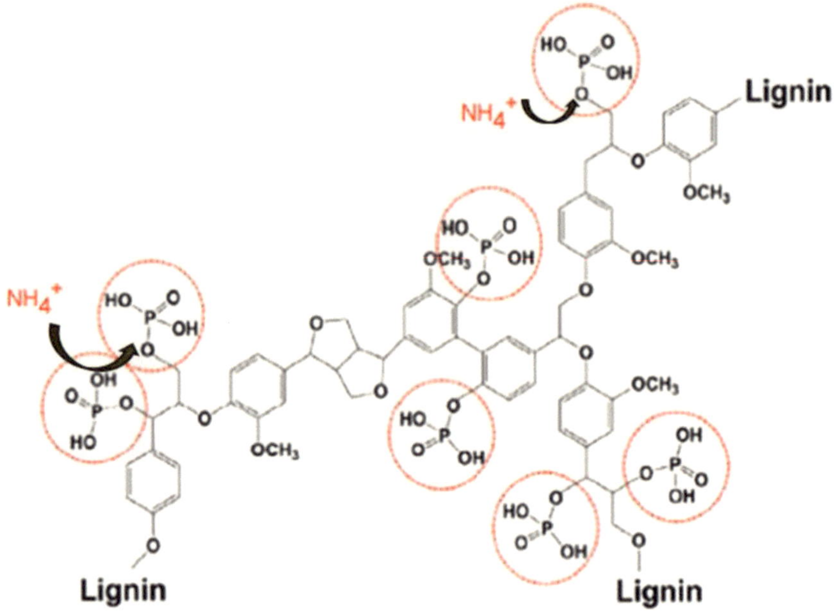

Figure 5.
Illustration of incorporation of phosphor and nitrogen in lignin structure.

3.1.2 Antibacterial activity of lignin

The antimicrobial property of biopolymers has been shown literature to depend on several factors, such as molecular mass, concentration, ability to be fixed into the wood structure and electrical charge [66]. Biopolymers with higher molecular masses have low biocidal activity, compared to high molecular mass biopolymers [67]. Literature studies shown that the antimicrobial activities of lignin can be inhibited by the presence phenolic monomers in lignin [68]. The lignin's antimicrobial activity depends on biomass source, the presence of hydroxyl and methoxy groups, and the extraction methods as follows: softwood organosolv > softwood kraft > grass organosolv due to the effect of acid-soluble lignin content [69]. Lignin as an antimicrobial agent is being used in commodity products like in plastic production [70], textile [71–73], medical materials, pest control, and healthcare products [74]. Lignin's and lignin nanoparticles chemical modification and combination with metals, for example Cu-lignin combination, have been shown to increase antimicrobial activity [75]. Thus, the use lignin as an antibacterial agent is believed to be a high value approach for lignin valorization.

4. Conclusions

Biopolymers are promising bio-based alternative to be used as biocides or barrier-forming compounds into the structure of wood. Designing a coating system with better performance on wood depends on understanding the interaction among individual wood constituents with the coating components. By using biopolymers as coating additives to protect the wood has environmental benefits and avoids the issue of wood recycling. Lignin based coating shows great potential in the future as a promising alternative to fossil-based polymers.

Acknowledgements

The author wishes to acknowledge the financial support from the Research Council of Norway (FME Centre for Environmentally friendly Energy Research Bio4Fuel, project number 25622).

Conflict of interest

The authors declare no conflict of interest.

Author details

Mihaela Tanase-Opedal
RISE PFI AS, Trondheim, Norway

*Address all correspondence to: mihaela.tanase@rise-pfi.no

References

[1] William L. Improving the durability performance of engineered wood products: A study of Europe and North America, 2018 Gottstein fellowship report, Gold Coast. 2018

[2] Available from: https://cwc.ca/wp-content/uploads/2019/03/publications-casestudy Innovative ApplicationsEngineeredWood.pdf

[3] Yadav R, Kumar J. Engineered wood products as a sustainable construction material: A review. In: Engineered Wood Products for Construction. London, UK: IntechOpen Book; 2021. DOI: 10.5772/intechopen.99597

[4] Will the Skyscrapers of the Future be Made Out Of Wood? Available from: https://www.nationalgeographic.com/science/article/skyscrapers-of-the-future-will-be-made-out-of-wood

[5] Cunningham MF, Campbell JD, Fu Z, Bohling J, Leroux JG, Mabee W, Robert T. Future green chemistry and sustainability needs in polymeric coatings. Green Chemistry. 2019;21:4919-4926

[6] Oliver CD, Nassar NT, Lippke BR, McCarter JB. Carbon, fossil fuel and biodivesrsity mitigation with wood and forests 2014. Journal of Sustainable Forestry. 2014;**33**:248-275. DOI: 10.1080/10549811.2013.839386

[7] Wood coatings: only modest growth - News and insights for the European coatings industry (abtdgt.de)

[8] Betts WD. The properties and performance of coal-tar creosote as wood preservative. In: Thompson R, editor. The Chemistry of Wood Preservation. Amsterdam: Woodhead Publishing, Elsevier; 2005. pp. 117-135

[9] Bull DC. The chemistry of chromated copper arsenate II. Preservative wood-interactions. Wood Science and Technology. 2001;**34**:459-466

[10] Lu S, Hamerton I. Recent developments in the chemistry of halogen-free flame retardant polymers. Progress in Polymer Science. 2002;**27**:1661-1712

[11] Liang D, Zhu X, Dai P, Lu X, Guo H, Que H, et al. Preparation of a novel lignin-based flame retardant for epoxy resin. Materials Chemistry and Physics. 2021;**259**:124101. DOI: 10.1016/j.matchemphys.2020.124101

[12] Thakur VK, Thakur MK. Recent advances in graft copolymerization and applications of chitosan: A review. ACS Sustainable Chemistry & Engineering. 2014;**2**:2637-2652. DOI: 10.1021/sc500634p

[13] Yang HT, Yu B, Xu XD, Bourbigot S, Wang H, Song PA. Lignin-derived bio-based flame retardants toward high-performance sustainable polymeric materials. Green Chemistry. 2020;**22**:2129-2161. DOI: 10.1039/D0GC00449A

[14] Adam O, Badot PM, Degiorgi F, Crini G. Mixture toxicity assessment of wood preservative pesticides in the freshwater amphipod Gammarus pulex (L.). Ecotoxicology and Environmental Safety. 2009;**72**(2):441-449

[15] Lin LD, Chen YF, Wang SY, Tsai MJ. Leachability, metal corrosion, and termite resistance of wood treated with copper-based preservative. International Biodeterioration & Biodegradation. 2009;**63**(4):533-538

[16] Temiz A, Yildiz UC, Nilsson T. Effects of the wood preservatives on

mechanical properties of yellow pine (Pinus sylvestris L.) wood. Building and Environment. 2006;**41**(7):910-914

[17] Rowell MR editor. Chemical Modification of wood, Chapter 14. In: Handbook of Wood Chemistry and Wood Composites. Madison, WI: CRC Press LLC; 2005. pp. 381-420

[18] Teaca C-A, Rosu D, Mustata F, Rusu T, Rosu L, Rosca I, et al. Natural coatings for wood. BioReources. 2019;**14**(2):4873-4901

[19] Andok A, Jesuet MSG. Biodegradable chitosan coating for wood protection. IOP Conference Series: Environmental Earth Sciences. 1053, 012036

[20] Mazela B, Polus-Ratajczak I. Reduction of preservative leaching by animal proteins in wood. Holzforschung. 2003;**57**:593-596

[21] Lambourne R, Strivens T. Paint and surface coatings. Theory and Practice. 2000;**37**:1-18

[22] Alam M, Akram D, Sharmin E, Zafar F, Ahmad S. Vegetable oil based eco-friendly coating materials: A review article. Arabian Journal of Chemistry. 2014;**7**:469-479. DOI: 10.1016/j.arabjc.2013.12.023

[23] Gerardin P, Petric M, Petrissans M, Lambert J, J.J. Ehrhrardt evolution of wood surface free energy after heat treatment. Polymer Degradation and Stability. 2007;**92**:653-657

[24] Pizzi A. Wood products and green chemistry. Annals of Forest Science. 2016;**73**:185-206. DOI: 10.1007/s13595-014-0448-3

[25] Niaounakis M. Biopolymers Reuse, Recycling, and Disposal. Amsterdam, The Neherlands: Elsevier; 2013. pp. 1-75

[26] Heinrich LA. Future opportunities for bio-based adhesives - advantages beyond renewability. Green Chemistry. 2019;**21**:1866-1888

[27] Vroman I, Tighzert L. Biodegradable polymers. Materials. 2009;**2**:307-344

[28] Song X, Tang S, Chi X, Han G, Bai L, Shi QS, et al. Valorization of lignin from biorefinery: Colloidal lignin micro-Nanospheres as multifunctional bio-based fillers for waterborne wood coating enhancement. ACS Sustainable Chemistry & Engineering. 2022;**10**:11655-11665

[29] Patachia S, Croitoru C. Biopolymers for wood preservation. In: Biopolymers and Biotech Admixtures for Eco-Efficient Construction Materials. Elsevier Ltd.; 2016. pp. 305-332. DOI: 10.1016/B978-0-08-100214-8.00014-2

[30] Gonzalez-Laredo RF, Rosales-Castro M, Rocha-Guzman NE, Gallegos-Infante JA, Moreno-Jimenez MR, Karchesy JJ. Wood preservation using natural products. Madera y bosques. 2015;**21**:63-76. DOI: 10.21829/mub.2015.210427

[31] Rosu L, Varganici D-C, Mustata F, Rosu D, Rosca I, Rusu T. Epoxy coatings based on modified vegetable oils for wood surface protection against fungal degradation. ACS Applied Materials & Interfaces. 2020;**12**(12):14443-14458. DOI: 10.1021/acsami.0c00682

[32] Rowell MR. Understanding wood surface chemistry and approaches to wood modification: A review. Polymers. 2021;**13**:2558. DOI: 10.3390/polym13152558

[33] Farid T, Rafiq IM, Ali A, Tang W. Transforming wood as next-generation structural and functional materials for a sustainable future. EcoMat. 2022;**4**:12154. DOI: 10.1002/eom2.12154

[34] Nejad M, Cooper P. Exterior wood coatings. Chapter 6. In: Wood in Civil Engineering. InTechOpen; 2020. pp. 110-129. DOI: 10.5772/67170

[35] Singh AP, Singh T, Rickard C. Visualising impregnated chitosan in Pinus radiata early wood cells using light and scanning electron microscopy. Micron. 2010;**41**:263-267

[36] Freeman et al. Past, present, and future of the wood preservation industry. Forest Products Journal. 2003;**53**:8-15

[37] Singh T, Chittenden C, Singh AP, Franich R. Chitosan as a potential wood preservative. Wood Processing Newsletter. 2008;**42**:11-23

[38] Passialis C, Grigoriou A, Voulgaridis EV. Utilization of oleoresin and bark extractives from Pinus halepensis mill in wood products. Forêt méllilermnéenne. 1995;**16**(1):19-27

[39] Ding X, Richter D, Matuana L, Heiden P. Efficient one-pot synthesis and loading of self-assembled amphiphilic chitosan nanoparticles for low-leaching wood preservation. Carbohydrate Polymers. 2011;**86**(1):58-64

[40] Croitoru C, Patachia S, Lunguleasa A. A mild method of wood impregnation with biopolymers and resins using 1-ethyl-3-methylimidazolium chloride as carrier. Chemical Engineering Research and Design. 2015;**93**:257-268

[41] Islam NM, Rahman F, Das KA, Hixiroglu S. An overview of different types and potential of bio-based adhesives used for wood products. International Journal of Adhesion & Adhesives. 2022;**112**:102992

[42] Sreejaya MM, Jeeven SR, Ramanunni K, Pillai PN, Ramkumar K, Anuvinda P, et al. Lignin based organic coatings and their applications: A review. Materials Today: Proceedings. 2022;**60**:494-501

[43] Siahkamari M, Emmanuel S, Hodge BD, Nejad M. Lignin-glyoxal: A fully biobased formaldehyde-free wood adhesive for interior engineered wood products. ACS Sustainable Chemistry & Engineering. 2022;**10**:3430-3441. DOI: 10.1021/acssuschemeng.1c06843

[44] Ragauskas AJ et al. 2014 DOI: 10.1126/science.1246843,

[45] Grossman A, Vermerris W. 2019. DOI: 10.1016/j.copbio.2018.10.009

[46] Ferrari F, Striani R, Fico D, Mahbubul AM, Greco A, Corcione EC. An overview on wood waste valorizations of biopolymers and biocomposites: Definition, classification, production, properties and applications, review. Polymers. 2022;**14**(24):5519. DOI: 10.3390/polym14245519

[47] Henn AK, Forsman N, Zou T, Österberg M. Colloidal lignin particles and epoxies for bio-based, durable and multiresistant nanostructured coatings. ACS Applied Materials & Interfaces. 2021;**13**:34793-34806

[48] Over LC, Grau E, Grelier S, Meier MAR, Cramail H. Synthesis and characterization of epoxy thermosetting polymers from Glycidylated Organosolv lignin and bisphenol a. Macromolecular Chemistry and Physics. 2017;**218**(4):1600411

[49] Lintinen K, Xiao Y, Bangalore Ashok R, Leskinen T, Sakarinen E, Sipponen M, et al. Closed cycle production of concentrated and dry Redispersible colloidal lignin particles with a three solvent polarity

exchange method. Green Chemistry. 2018;**20**:843-850

[50] Bao W, Deng Z, Zhang S, Ji Z, Zhang H. Next-generation composite coating system: Nanocoating. Frontiers in Materials. 2019;**6**(72):1-6

[51] Forsman N, Lozhechnikova A, Khakalo A, Johansson LS, Vartiainen J, Österberg M. Layer-by-layer assembled hydrophobic coatings for cellulose Nanofibril films and textiles, made of Polylysine and natural wax particles. Carbohydrate Polymers. 2017;**173**:392-402

[52] Kosak Söz C, Yilgör E, Yilgör I. Influence of the average surface roughness on the formation of Superhydrophobic polymer surfaces through spin-coating with hydrophobic Fumed silica. Polymer. 2015;**62**:118-128

[53] Buss F, Roberts CC, Crawford KS, Peters K, Francis LF. Effect of soluble polymer binder on particle distribution in a drying particulate coating. Journal of Colloid and Interface Science. 2011;**359**:112-120

[54] De Francisco R, Tiemblo P, Hoyos M, González-Arellano C, García N, Berglund L, et al. Multipurpose ultra and Superhydrophobic surfaces based on Oligodimethylsiloxane-modified Nanosilica. ACS Applied Materials & Interfaces. 2014;**6**:18998-19010

[55] Das S, Kumar S, Samal SK, Mohanty S, Nayak SK. A review on Superhydrophobic polymer Nanocoatings: Recent development and applications. Industrial and Engineering Chemistry Research. 2018;**57**:2727-2745

[56] Bourbigot S, Le Bras M, Delobel R, Tremillon J-M. Synergistic effect of zeolite in an intumescence process. Study of the interactions between the polymer and the additives. Journal of the Chemical Society, Faraday Transactions. 1996;**92**:3435

[57] Gao C, Zhou L, Yao S, Qin C, Fatehi P. Phosphorylated Kraft lignin with improved thermal stability. International Journal of Biological Macromolecules. 2020;**162**:1642-1652

[58] Song P, Cao Z, Fu S, Fang Z, Wu Q, Ye J. Thermal degradation and flamr retardarncy properties of ABS/lignin: Effects of lignin content and reactive compatibilization. Thermochimica Acta. 2011;**518**:59-65

[59] Dai P, Liang M, Ma X, Luo Y, He M, Gu X, et al. Highly efficient, environmentally friendly lignin-based flame retardant used in epoxy resin. ACS Omega. 2020;**5**:32084-32093

[60] Liu L, Huang G, Song P, Yu Y, Fu S. Converting industrial alkali lignin to biobased functional additives for improving fire behavior and smoke suppression of polybutylene succinate. ACS Sustainable Chemistry & Engineering. 2016;**4**:4732-4742

[61] Costes L, Laoutid F, Aguedo M, Richel A, Brohez S, Delvosalle C, et al. Phosphorus and nitrogen derivatization as efficient route for improvement of lignin flame retardant action in PLA. European Polymer Journal. 2016;**84**:652-667

[62] Wu W, He H, Liu T, Wei R, Cao X, Sun Q, et al. Synergetic enhancement on flame retardancy by melamine phosphate modified lignin in rice husk ash filled P34HB biocomposites. Composites Science and Technology. 2018;**168**:246-254

[63] Liu L, Qian M, Song PA, Huang G, Yu Y, Fu S. Fabrication of green lignin-based flame retardants for enhancing the

thermal and fire retardancy properties of polypropylene/wood composites. ACS Sustainable Chemistry & Engineering. 2016;**4**:2422-2431

[64] Zhang R, Xiao X, Tai Q, Huang H, Hu Y. Modification of lignin and its application as char agent in itumecent flame retardant poly(lactic acid). Polymer Engineering and Science. 2012;**52**(12):2620-2626. DOI: 10.1002/pen.23214

[65] Collet B, Lopez-Cuesta J-M, Laoutid F, Fery L. Lignin nanoparticles as a promising way for enhancing lignin flame retardant effect in polylactide. Materials. 2019;**12**:2132. DOI: 10.3390/ma12132132

[66] Goy RC, de Britto D, Assis OBG. A review of the antimicrobial activity of chitosan. Polímeros. 2009;**19**(3):241-247

[67] Badawy MEI, Rabea EI. A biopolymer chitosan and its derivatives as promising antimicrobial agents against plant pathogens and their applications in crop protection. International Journal of Carbohydrate Chemistry. 2011;**3**:100-129

[68] Jung HG, Fahey GC. Nutritional implications of phenolic monomers and lignin: A review. Journal of Animal Science. 1983;**57**(1):206-219

[69] Alzagameem A, Klein ES, Bergs M, Tung DX, Korte I, Dohlen S, et al. Antimicrobial activity of lignin and lignin-derived cellulose and chitosan composites against selected pathogenic and spoilage microorganisms. Polymers. 2019;**11**:670. DOI: 10.3390/polym11040670

[70] Klein A, Rumpf K, Kreyenschmidt, and Schulze. Antimicrobial activity of LigninDerived polyurethane coatings prepared from unmodified and Demethylated Lignins. Coatings. 2019;**9**(8):494. DOI: 10.3390/coatings9080494

[71] Sriroth K, Sunthornvarabhas J. Lignin from sugar process as natural antimicrobial agent. Biochemistry and Pharmacology: Open Access. 2018;**07**(01):1-4. DOI: 10.4172/2167-0501.1000239

[72] Sunthornvarabhas J, Liengprayoon S, Suwonsichon T. Antimicrobial kinetic activities of lignin from sugarcane bagasse for textile product. Industrial Crops and Products. 2017;**109**:857-861. DOI: 10.1016/j.indcrop.2017.09.059

[73] Sunthornvarabhas J, Liengprayoon S, Lerksamran T, Buratcharin C, Suwonsichon TI, Vanichsriratana W, et al. Utilization of lignin extracts from sugarcane bagasse as bio-based antimicrobial fabrics. Sugar Tech. 2018;**21**(2):355-363. DOI: 10.1007/s12355-018-0683-2

[74] Gordobil O, Herrera R, Yahyaoui M, İlk S, Kaya M, Labidi J. Potential use of Kraft and Organosolv Lignins as a natural additive for healthcare products. RSC Advances. 2018;**8**(43):24525-24533. DOI: 10.1039/c8ra02255k

[75] Sinisi V, Pelagatti P, Carcelli M, Migliori A, Mantovani L, Righi L, et al. A green approach to copper-containing pesticides: Antimicrobial and antifungal activity of Brochantite supported on lignin for the development of biobased plant protection products. ACS Sustainable Chemistry & Engineering. 2018;**7**(3):3213-3221. DOI: 10.1021/acssuschemeng.8b05135